Suzuki GS1000 Fours Owners Workshop Manual

by Martyn Meek

Models covered:
GS1000 C 997 cc Introduced into USA only, October 1977
GS1000 EC 997 cc Introduced USA, October 1977, UK, May 1978
GS1000 HC 997 cc Introduced into UK only, February 1978
GS1000 N 997 cc Introduced into USA only, 1978
GS1000 EN 997 cc Introduced into USA and UK, 1979
GS1000 L 997 cc Introduced into USA only, 1979
GS1000 S 997 cc Introduced into USA and UK, 1979

ISBN 978 0 85696 484 8

ABCDE
FGHIJ
KLMNO

2

Printed in the UK *(484-9Q4)*

Haynes Publishing Group
Sparkford Nr Yeovil
Somerset BA22 7JJ England

Haynes Publications, Inc
859 Lawrence Drive
Newbury Park
California 91320 USA

Acknowledgements

Our thanks are due to P.R. Taylor and Sons of Chippenham who loaned the GS1000 HC featured in the photographs throughout this manual. Our thanks are also due to Heron Suzuki GB Limited for permission to reproduce their drawings, and to members of the Technical Service Department of that company, who gave much valuable advice and checked the content of this manual, suggesting ways in which the text could be improved.

Alan Jackson supervised and assisted with the dismantling of the machine and also the rebuilding sequences, and devised various ingenious methods for overcoming the lack of service tools. Tony Steadman arranged and took the photographs and Mansur Darlington edited the text.

We should also like to thank the Avon Rubber Company who kindly supplied us with information and advice about tyre fitting, and NGK Spark Plugs (UK) Limited, for information and photographs relating to sparking plug conditions.

About this manual

The purpose of this manual is to present the owner with a concise and graphic guide which will enable him to tackle any operation from basic routine maintenance to a major overhaul. It has been assumed that any work would be undertaken without the luxury of a well-equipped workshop and a range of manufacturer's service tools.

To this end, the machine featured in the manual was stripped and rebuilt in our own workshop, by a team comprising a mechanic, a photographer and the author. The resulting photographic sequence depicts events as they took place, the hands shown being those of the author and the mechanic.

The use of specialised, and expensive, service tools was avoided unless their use was considered to be essential due to risk of breakage or injury. There is usually some way of improvising a method of removing a stubborn component, provided that a suitable degree of care is exercised.

The author learnt his motorcycle mechanics over a number of years, faced with the same difficulties and using similar facilities to those encountered by most owners. It is hoped that this practical experience can be passed on through the pages of this manual.

Where possible, a well-used example of the machine is chosen for the workshop project, as this highlights any areas which might be particularly prone to giving rise to problems. In this way, any such difficulties are encountered and resolved before the text is written, and the techniques used to deal with them can be incorporated in the relevant section. Armed with a working knowledge of the machine, the author undertakes a considerable amount of research in order that the maximum amount of data can be included in this manual.

Each Chapter is divided into numbered sections. Within these sections are numbered paragraphs. Cross reference throughout the manual is quite straightforward and logical. When reference is made 'See Section 6.10' it means Section 6, paragraph 10 in the same Chapter. If another Chapter were intended the reference would read, for example, 'See Chapter 2, Section 6.10'. All the photographs are captioned with a section/paragraph number to which they refer and are relevant to the Chapter text adjacent.

Figures (usually line illustrations) appear in a logical but numerical order, within a given Chapter. Fig. 1.1 therefore refers to the first figure in Chapter 1.

Left-hand and right-hand descriptions of the machines and their components refer to the left and right of a given machine when the rider is seated normally.

Motorcycle manufacturers continually make changes to specifications and recommendations, and these, when notified, are incorporated into our manuals at the earliest opportunity.

Whilst every care is taken to ensure that the information in this manual is correct no liability can be accepted by the author or publishers for loss, damage or injury caused by any errors in or omissions from the information given.

Contents

Note: General description and specifications are given in each Chapter immediately after the list of contents. Fault diagnosis is given at the end of the Chapter.

Left-hand view of Suzuki GS1000 HC

Introduction to the Suzuki GS1000 models

Although the Suzuki Motor Company Limited commenced manufacturing motorcycles as early as 1936, it was not until 1963 that their machines were first imported into the UK. The first of the twin cylinder models, the T10, became available during 1964, and it was immediately obvious that this particular model would be well-received by holders of a provisional driving license, who are restricted to an engine capacity limit of 250 cc. Its popularity was mostly due to its impressive performance, in terms of both speed and fuel economy, and its advanced specification, including an electric starter and a hydraulic rear brake.

The number of models available from Suzuki has increased steadily; in 1971, one could choose a model ranging in capacity from 50 cc to 750 cc; in the last two or three years, in common with the other large Japanese companies, their range of models increased quite dramatically. There is now an even larger choice of machines available, with engine capacities ranging still from 50 cc, but now up to 1100 cc, and encompassing machine types as varied as competition only, Moto-cross, single-cylinder two-strokes, and high performance four-cylinder, four-stroke road machines of prestigious specifications.

The GS1000 is one of the latter type of machine and marks the entry of Suzuki into the World's 'Superbike' market which has, of late, changed from consisting mainly of 750 cc machines to those in the 900 cc to 1300 cc bracket. The GS 1000, design is very similar to, and indeed mechanically based upon, the existing models in the four-cylinder GS range, namely the GS550 and 750. When introduced in 1978, however, the design broke certain new ground. The major departure from accepted practice is in the suspension and frame departments.

The front forks operate on the normal hydraulic telescopic fork principle, but derive extra damping by variable air assistance. The rear suspension, is, again, basically conventional, a swinging arm with coil springing/oil damping (or air adjustable damping) suspension units. The normal oil damped units however, incorporate a sophisticated variable rebound damping adjustment facility. The combination of this suspension and a sturdy, non-flexing frame meant the GS1000 handled extremely well, and ensured it was an instant success coming, as it did, after a long succession of Japanese machines which were incapable of reproducing traditional 'European-type' handling characteristics.

The initial C and HC models were supplemented later in 1978 by the EC model. In 1979, the N and EN models were introduced, being basically the same machines as their immediate predecessors, with a change of paint scheme, and detail refinements and modifications. The range of current models has now increased to four; the new additions being the L and S models. The L variant is styled in the 'semi-Chopper' mould; a style becoming increasingly popular and now involving all the four large Japanese manufacturers, and is, at present, only available in the U.S.A. The S model, however, is designed more as a sports machine, and was originally planned for the European market, coming as it does, with increased power output, lower handlebars, and a small sports fairing. A limited number only of the S models have been allocated to the US market.

Reference is made throughout the text, in the dismantling and reassembly procedures, to the various differences in each model's equipment.

Model dimension and weights

	GS1000L	GS1000S	All other models
Overall length	2265 mm (89.2 in)	2225 mm (87.6 in)	2225 mm (87.6 in)
Overall width	890 mm (35.0 in)	735 mm (28.9 in)	850 mm (33.5 in)
Overall height	1245 mm (49.0 in)	1255 mm (49.4)	1165 mm (45.9 in)
Wheelbase	1535 mm (60.4 in)	1505 mm (59.3 in)	1505 mm (59.3 in)
Ground clearance	160 mm (6.3 in)	160 mm (6.3 in)	155 mm (6.1 in)
Dry weight	240 kg (529 lb)	238 kg (525 lb)	230 kg (507 lb) (E model 234 kg [516 lb])

Ordering spare parts

When ordering spare parts for any Suzuki, it is advisable to deal direct with an official Suzuki agent who should be able to supply most of the parts ex stock. Parts cannot be obtained from Suzuki direct and all orders must be routed via an approved agent even if the parts required are not held in stock. Always, quote the engine and frame numbers in full, especially if parts are required for earlier models.

The frame and engine numbers are stamped on a Manufacturer's Plate riveted to the steering head on the left-hand side. The frame number is also stamped on the frame itself on the right-hand side of the steering head. The engine number is stamped on the upper crankcase.

Use only genuine Suzuki spares. Some pattern parts are available that are made in Japan and may be packed in similar looking packages. They should only be used if genuine parts are hard to obtain or in an emergency, for they do not normally last as long as genuine parts, even though there may be a price advantage.

Some of the more expendable parts such as spark plugs, bulbs, tyres, oils and greases etc., can be obtained from accessory shops and motor factors, who have convenient opening hours, and can often be found not far from home. It is also possible to obtain parts on a Mail Order basis from a number of specialists who advertise regularly in the motorcycle magazines.

Location of: Frame number

Location of: Engine number

Safety first!

Professional motor mechanics are trained in safe working procedures. However enthusiastic you may be about getting on with the job in hand, do take the time to ensure that your safety is not put at risk. A moment's lack of attention can result in an accident, as can failure to observe certain elementary precautions.

There will always be new ways of having accidents, and the following points do not pretend to be a comprehensive list of all dangers; they are intended rather to make you aware of the risks and to encourage a safety-conscious approach to all work you carry out on your vehicle.

Essential DOs and DON'Ts

DON'T start the engine without first ascertaining that the transmission is in neutral.

DON'T suddenly remove the filler cap from a hot cooling system – cover it with a cloth and release the pressure gradually first, or you may get scalded by escaping coolant.

DON'T attempt to drain oil until you are sure it has cooled sufficiently to avoid scalding you.

DON'T grasp any part of the engine, exhaust or silencer without first ascertaining that it is sufficiently cool to avoid burning you.

DON'T allow brake fluid or antifreeze to contact the machine's paintwork or plastic components.

DON'T syphon toxic liquids such as fuel, brake fluid or antifreeze by mouth, or allow them to remain on your skin.

DON'T inhale dust – it may be injurious to health (see *Asbestos* heading).

DON'T allow any spilt oil or grease to remain on the floor – wipe it up straight away, before someone slips on it.

DON'T use ill-fitting spanners or other tools which may slip and cause injury.

DON'T attempt to lift a heavy component which may be beyond your capability – get assistance.

DON'T rush to finish a job, or take unverified short cuts.

DON'T allow children or animals in or around an unattended vehicle.

DON'T inflate a tyre to a pressure above the recommended maximum. Apart from overstressing the carcase and wheel rim, in extreme cases the tyre may blow off forcibly.

DO ensure that the machine is supported securely at all times. This is especially important when the machine is blocked up to aid wheel or fork removal.

DO take care when attempting to slacken a stubborn nut or bolt. It is generally better to pull on a spanner, rather than push, so that if slippage occurs you fall away from the machine rather than on to it.

DO wear eye protection when using power tools such as drill, sander, bench grinder etc.

DO use a barrier cream on your hands prior to undertaking dirty jobs – it will protect your skin from infection as well as making the dirt easier to remove afterwards; but make sure your hands aren't left slippery. Note that long-term contact with used engine oil can be a health hazard.

DO keep loose clothing (cuffs, tie etc) and long hair well out of the way of moving mechanical parts.

DO remove rings, wristwatch etc, before working on the vehicle – especially the electrical system.

DO keep your work area tidy – it is only too easy to fall over articles left lying around.

DO exercise caution when compressing springs for removal or installation. Ensure that the tension is applied and released in a controlled manner, using suitable tools which preclude the possibility of the spring escaping violently.

DO ensure that any lifting tackle used has a safe working load rating adequate for the job.

DO get someone to check periodically that all is well, when working alone on the vehicle.

DO carry out work in a logical sequence and check that everything is correctly assembled and tightened afterwards.

DO remember that your vehicle's safety affects that of yourself and others. If in doubt on any point, get specialist advice.

IF, in spite of following these precautions, you are unfortunate enough to injure yourself, seek medical attention as soon as possible.

Asbestos

Certain friction, insulating, sealing, and other products - such as brake linings, clutch linings, gaskets, etc – contain asbestos. *Extreme care must be taken to avoid inhalation of dust from such products since it is hazardous to health.* If in doubt, assume that they *do* contain asbestos.

Fire

Remember at all times that petrol (gasoline) is highly flammable. Never smoke, or have any kind of naked flame around, when working on the vehicle. But the risk does not end there – a spark caused by an electrical short-circuit, by two metal surfaces contacting each other, by careless use of tools, or even by static electricity built up in your body under certain conditions, can ignite petrol vapour, which in a confined space is highly explosive.

Always disconnect the battery earth (ground) terminal before working on any part of the fuel or electrical system, and never risk spilling fuel on to a hot engine or exhaust.

It is recommended that a fire extinguisher of a type suitable for fuel and electrical fires is kept handy in the garage or workplace at all times. Never try to extinguish a fuel or electrical fire with water.

Note: *Any reference to a 'torch' appearing in this manual should always be taken to mean a hand-held battery-operated electric lamp or flashlight. It does **not** mean a welding/gas torch or blowlamp.*

Fumes

Certain fumes are highly toxic and can quickly cause unconsciousness and even death if inhaled to any extent. Petrol (gasoline) vapour comes into this category, as do the vapours from certain solvents such as trichloroethylene. Any draining or pouring of such volatile fluids should be done in a well ventilated area.

When using cleaning fluids and solvents, read the instructions carefully. Never use materials from unmarked containers – they may give off poisonous vapours.

Never run the engine of a motor vehicle in an enclosed space such as a garage. Exhaust fumes contain carbon monoxide which is extremely poisonous; if you need to run the engine, always do so in the open air or at least have the rear of the vehicle outside the workplace.

The battery

Never cause a spark, or allow a naked light, near the vehicle's battery. It will normally be giving off a certain amount of hydrogen gas, which is highly explosive.

Always disconnect the battery earth (ground) terminal before working on the fuel or electrical systems.

If possible, loosen the filler plugs or cover when charging the battery from an external source. Do not charge at an excessive rate or the battery may burst.

Take care when topping up and when carrying the battery. The acid electrolyte, even when diluted, is very corrosive and should not be allowed to contact the eyes or skin.

If you ever need to prepare electrolyte yourself, always add the acid slowly to the water, and never the other way round. Protect against splashes by wearing rubber gloves and goggles.

Mains electricity and electrical equipment

When using an electric power tool, inspection light etc, always ensure that the appliance is correctly connected to its plug and that, where necessary, it is properly earthed (grounded). Do not use such appliances in damp conditions and, again, beware of creating a spark or applying excessive heat in the vicinity of fuel or fuel vapour. Also ensure that the appliances meet the relevant national safety standards.

Ignition HT voltage

A severe electric shock can result from touching certain parts of the ignition system, such as the HT leads, when the engine is running or being cranked, particularly if components are damp or the insulation is defective. Where an electronic ignition system is fitted, the HT voltage is much higher and could prove fatal.

Routine maintenance

Periodic routine maintenance is a continuous process that commences immediately the machine is used and continues until the machine is no longer fit for service. It must be carried out at specified mileage recordings or on a calendar basis if the machine is not used regularly, whichever is the sooner. Maintenance should be regarded as an insurance policy, to help keep the machine in the peak of condition and to ensure long, trouble-free service. It has the additional benefit of giving early warning of any faults that may develop and will act as a safety check, to the obvious advantage of both rider and machine alike.

The various maintenance tasks are described under their respective mileage and calendar headings. Accompanying photos or diagrams are provided, where necessary. It should be remembered that the interval between the various maintenance tasks serves only as a guide. As the machine gets older, is driven hard, or is used under particularly adverse conditions, it is advisable to reduce the period between each check.

For ease of reference each service operation is described in detail under the relevant heading. However, if further general information is required it can be found within the manual in the relevant Chapter.

Although no special tools are required for routine maintenance, a good selection of general workshop tools are essential. Included in the tools must be a range of metric ring or combination spanners, a selection of crosshead screwdrivers, and two pairs of circlip pliers, one external opening and the other internal opening. Additionally, owing to the extreme tightness of most casing screws on Japanese machines, an impact screwdriver, together with a choice of large or small cross-head screw bits, is absolutely indispensable. This is particularly so if the engine has not been dismantled since leaving the factory.

Weekly or every 250 miles

1 Tyre pressures

Check the tyre pressures with a pressure gauge that is known to be accurate. Always check the pressures when the tyres are cold. If the tyres are checked after the machine has travelled a number of miles, the tyres will have become hot and consequently the pressure will have increased, possibly as much as 8 psi. A false reading will therefore always result.

Tyres pressures:	Solo	Pillion
Front tyre	25 psi (1.75 kg cm²)	28 psi (2.00 kg cm²)
Rear tyre	28 psi (2.00 kg cm²)	32 psi (2.25 kg cm²)

For continuous high speed riding, the pressures should be increased to:

	Solo	Pillion
Front tyre	28 psi (2.00 kg cm²)	32 psi (2.25 kg cm²)
Rear tyre	36 psi (2.50 kg cm²)	40 psi (2.80 kg cm²)

2 Engine oil level

Place the machine on the centre stand and by viewing the sight-glass in the primary drive casing, check that the engine/transmission oil level is between the two level marks. The machine must be upright because even a slight lean will give a false reading. If necessary, replenish the engine with the correct quantity of SAE 10W/40 engine oil. The filler cap is situated in the top of the primary drive cover.

Check engine/transmission oil level through sight-glass

Replenish through aperture in top of casing

3 Control cable lubrication

Apply a few drops of motor oil to the exposed inner portion of each control cable. This will prevent drying-up of the cables between the more thorough lubrication that should be carried out during the 3000 mile/4 monthly service.

4 Safety check

Give the machine a close visual inspection, checking for loose nuts and fittings, frayed control cables etc. Check the tyres for damage, especially splitting of the sidewalls. Remove any stones or other objects caught between the treads. This is particularly important on the front tyre, where rapid deflation due to penetration of the inner tube will almost certainly cause total loss of control. When checking the tyres for damage, they should be examined for tread depth in view of both the legal and safety aspects. It is vital to keep the tread depth within the legal limits of 1 mm of depth over three-quarters of the tread breadth and around the entire circumference. Many riders, however, consider nearer 2 mm to be the limit for secure road-holding, traction and braking, especially when riding under adverse weather and road conditions. Suzuki recommend replacement of the front tyre when there is 1.6 mm (0.06 in) of tread remaining, and replacement of the rear tyre when wear has reached the 2.0 mm (0.08 in) level.

5 Legal check

Ensure that the lights, horn, and flashing indicators all function correctly, and check that the speedometer is working and is reasonably accurate; the law requires it to be accurate within 10 per cent.

Monthly or every 1000 miles

Complete the tasks listed under the weekly 250 mile heading and then carry out the following checks:

1 Hydraulic fluid level

Check the level of the hydraulic fluid in the front brake master cylinder reservoir, on the handlebars, and also the rear brake reservoir, behind the right-hand frame side cover. The level can be seen through the transparent reservoir and should be between the upper and lower level marks. Ensure that the handlebars are in the central position when a level reading is taken from the front reservoir, and also when the cap and diaphragm are removed. During normal service, it is unlikely that the hydraulic fluid level will fall dramatically unless a leak has developed in the system. If this occurs, the fault should be remedied **AT ONCE**. The level will fall slowly as the brake pads wear, and the fluid deficiency should be corrected when required. Always use a hydraulic fluid of DOT 3 or SAE J1703 specification and if possible do not mix different types of fluid, even if the specifications appear the same. This will prevent the possibility of two incompatible fluids being mixed and the resultant chemical reactions damaging the seals.

If the level in either reservoir has been allowed to fall below the specified limit, and air has entered the system, the brake in question must be bled, as described in Chapter 5, Section 7.

2 Final drive chain lubrication

In order that final drive chain life be extended as much as possible, regular lubrication and adjustment is essential. This is particularly so when the chain is not enclosed or is fitted to a machine transmitting high power to the rear wheel. The chain may be lubricated whilst it is in situ on the machine by the application of a heavy oil or grease. Ordinary engine oil can be used, though owing to the speed with which it is flung off the rotating chain, its effective life is limited. Do not use aerosol chain lubricant as this may damage the sealing O-rings causing early chain failure.

It is strongly recommended that the chain is cleaned using paraffin before the application of the lubricant. This will remove any road grit which would otherwise become sealed in and accelerate wear of the rollers and sprocket teeth. The use of petrol or other solvents is not recommended as these too may damage the O-rings.

3 Final drive chain adjustment

Check the slack in the final drive chain. The correct up and down movement, as measured at the mid-point of the chain lower run, should be 20 mm (0.8 in). Adjustment should be carried out as follows. Place the machine on the centre stand so that the rear wheel is clear of the ground and free to rotate. Remove the split pin from the wheel spindle and slacken the wheel nut a few turns. Loosen the locknuts on the two chain adjuster bolts, and slacken off the brake torque rod nuts.

Rotation of the adjuster bolts in a clockwise direction will tighten the chain. Tighten each bolt a similar number of turns so that wheel alignment is maintained. This can be verified by checking that the mark on the outer face of each chain adjuster is aligned with the same aligning mark on each fork end. With the adjustment correct, tighten the wheel nut and fit a new split pin. Finally, retighten the adjuster bolt locknuts, and tighten the torque rod nuts, securing them by means of the split pins.

4 Battery electrolyte level

Maintenance of the battery fitted to the GS1000 range is normally limited to keeping the electrolyte level correct. The transparent plastic case of the battery permits the level of the electrolyte to be observed when the right-hand frame sidecover is detached. The battery is retained in a metal carrier below the air cleaner element and case, which in turn, is situated below the dualseat. To replenish the battery, is must be removed from the machine. Raise the dualseat, remove the air cleaner case and disconnect the battery leads and retaining straps; the battery may now be lifted out. Note that a thin vent pipe is fitted to the battery case; this must always be kept free from blockage, and must be routed so as to avoid any sharp turns or 'kinks'. The electrolyte solution should be between the Upper and Lower level lines. If the solution is low it should be replenished, using distilled water. The lead plates and their separators can be seen through the transparent case, a further guide to the general condition of the battery.

Unless acid is spilt, as may occur if the machine falls over, the electrolyte should always be topped up with distilled water, to restore the correct level. If acid is spilt on any part of the machine, it should be neutralised with an alkali such as washing soda and washed away with plenty of water, otherwise serious corrosion will occur. Top up with sulphuric acid of the correct specific gravity 1.260 - 1.280 only when spillage has occurred.

Maintain electrolyte level between upper and lower lines

Two monthly or every 1500 miles

Complete the checks listed under the weekly/250 mile and monthly/1000 mile headings and then carry out the following:

1 Change the engine oil

The oil should be changed with the engine at its normal operating temperature, preferably after a run. This ensures that the oil is relatively thin and will drain more quickly and completely.

Drain the engine oil by removing the drain plug from the under side of the sump and the filler cap from the top of the primary drive cover. Unscrew also the oil filter chamber drain plug to allow the small amount of lubricant within to escape. Ensure that a container of sufficient size is placed below the engine to catch the oil. When all the oil has drained off, refit and tighten the drain plugs, after checking that their sealing washers are in good condition.

Replenish the engine with approximately 3.4 lit (7.2/6.0 US/Imp pint) of SAE 10W/40 motor oil. Allow the oil to settle, and run the engine briefly at idling speed. Then recheck the oil level, by means of the sight-glass, and, if necessary, add more lubricant. The oil should be roughly three-quarters of the way up the sight-glass.

2 Clean the air filter element

The air filter element is fitted in a case located below the dualseat. The plastic element case has a steel lid which when removed allows access to the air filter element. Raise the dualseat and remove the air filter case lid, which is hinged at one end and retained by one screw at the other. Pull the air filter element upwards at its forward end, releasing the tension on the spring bracket that holds the element in position.

The element is of the dry paper cartridge type and should be blown clean by the use of compressed air. If a source of high pressure air is not available, then the use of a tyre pump makes an acceptable substitute. Ensure that the air is blown through the element from the inside. If the air is blown onto the outside surface of the element, then the dust particles will be forced further into the element, causing more, rather than less, restriction due to clogging. If there are any signs of oil, excessive dirt, or actual break-up of the element, then it must be replaced immediately.

Reinstall the element by reversing the removal procedure, ensuring that the spring bracket on the filter element forward end is located correctly with the securing bracket within the element case.

Four monthly or every 3000 miles

Carry out the tasks described in the weekly, monthly, and two monthly sections and then carry out the following:

1 Oil filter renewal

The oil filter element should be renewed at every second oil change. After draining the engine oil, remove the separate oil filler drain plug situated at the bottom front face of the oil filler chamber cover, and drain the remaining oil. Ensure a container is placed below the chamber to catch the escaping oil. To release the chamber cover, remove the three domed nuts. The cover is under tension from the filter locating spring and so may fly off if care is not taken. Lift out the spring and oil filter element. No attempt should be made to clean the old filter; it must be discarded and a new component fitted. Clean the filter chamber before inserting the new element which should be fitted with the rubber seal end facing inwards. Check the condition of the chamber cover sealing ring before replacing the cover.

Note than when the engine oil and filter are both replaced at the same service interval, 3.8 lit (8.0/6.6 US/Imp pint) of SAE 10 W/40 engine oil will be needed to fully replenish the sump.

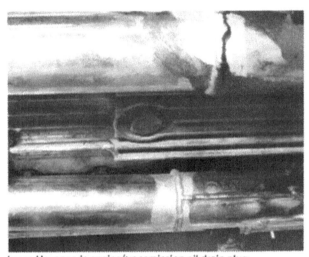

Large Hexagon is engine/transmission oil drain plug

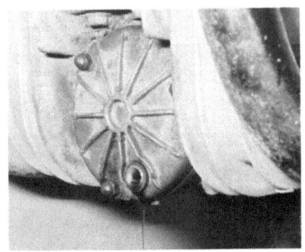

Remove drain bolt to allow oil from filter housing to escape

Remove cover plate for access to filter element

2 Cleaning and adjusting the contact breaker points

To gain access to the contact breaker assembly it is necessary to remove the engine right-hand cover which is retained by three screws.

Before adjusting the points, examine each set for burning or pitting. Clean or renew the points as necessary. See Section 5 of Chapter 3. Deposits due to arcing can be removed while the contact breaker unit is in situ on the machine, using a very fine Swiss file or emery paper (No.400) backed by a thin strip of tin. If the pitting or burning is excessive, the contact breaker unit in question must be removed for points dressing or renewal.

The points are marked '1.4' and '2.3' adjacent to the relevant contact set, indicating the pair of cylinders they serve. Set the gap of points marked 1.4 first. Turn the crankshaft using a spanner on the large hexagonal washer securing the cam, until these points are fully open. Measure the gap with a feeler gauge, and adjust if necessary. Standard gap: 0.3 – 0.4 mm (0.012 – 0.016 inch).

If the gap requires adjustment, slacken slightly the slotted screw which secures the fixed contact. A screwdriver should be engaged between the slot in the fixed contact, and the two pins on the contact breaker back plate; by turning the screwdriver the gap may be opened or closed. Tighten the screw and recheck the gap.

Turn the crankshaft so that the points marked 2.3 are fully open, and repeat the procedure above. Do not slacken the two screws which secure the 2.3 contact set base plate to the main base plate; this will upset the timing. It is important that both points should be set to the same gap, as the gap determines the moment when the contacts open, and thus the ignition timing.

3 Checking and resetting the ignition timing

Whenever the contact breaker unit receives attention, the ignition timing should be checked as a matter of course and adjusted, if necessary.

Apply a spanner to the engine turning hexagon and turn the engine in a forward direction, whilst viewing the automatic timing unit through the inspection aperture in the contact breaker stator plate. It will be seen that there is a set of three scribed lines on each side of the ATU.

Commence ignition timing on the left-hand contact breaker set, which controls cylinders No. 1 and 4. To determine the point at which the points open, connect a 12 volt bulb between the moving point and a suitable earth point on the engine. With the ignition turned on, the bulb will illuminate when the points are open. Rotate the engine until the 'F1-4' mark on the ATU is in **exact** alignment with the index pointer mark on the plate fitted to the rear of the stator plate. If the ignition is correct, the points should be on the verge of opening when this position is reached. This will be indicated by the flickering of the bulb. To adjust the ignition timing on No. 1 and 4 cylinders, slacken the three screws which pass through the elongated holes in the stator plate periphery. Rotate the plate until the light flickers and then tighten the screw. Turn the engine backwards 90° and then forwards again to check the setting.

Check the ignition timing on No.2 and 3 cylinders in a similar manner, by connecting the bulb to the right-hand contact breaker set and referring to the F2-3 mark on the ATU. If the timing is incorrect, slacken the two screws which hold the right-hand contact breaker assembly mounting plate to the main stator plate. Move the plate to the correct position and tighten the screw. Recheck the timing on No. 2 and 3 cylinders.

The ignition timing may also be checked by the use of a stroboscope when the engine is running.

Check and adjust the 1 – 4 contact breaker first and then the 2 – 3 contact breaker. At 1500 rpm and below, the F mark should align with the index mark. Full advance should be reached at 2500 rpm, when the unmarked advance line to the right of the F mark should be in line with the index mark. Adjustment should be made in the same manner as described for manual ignition timing.

Before replacing the contact breaker cover, apply a small quantity of light oil or grease to the cam lubricating wick. Do not overlubricate, as excess oil may find its way onto the points, causing ignition failure.

4 Valve clearance checking and adjustment

To gain access to the camshafts and cam followers the petrol tank must be removed as described in Chapter 2, Section 2, and the camshaft cover detached. After removal of the cover bolts, the seal between the cover and the gasket may be broken by the judicious use of a rawhide mallet. Strike only those parts of the cover which are well supported by lugs.

Unscrew the spark plugs and remove the contact breaker cover from the right-hand side of the engine. The clearance between each cam and cam follower must be checked and if necessary adjusted by removal of the existing adjuster pad and replacement of a pad of suitable thickness. Make the clearance check and adjustment of each valve in sequence and then continue with the next valve. As shown in the accompanying diagram, both operations should be carried out with the cam lobe in question placed in one or two alternative positions. Rotate the engine in a forward direction by means of the engine turning hexagon on the contact breaker cam end. Use only the 19 mm hexagon for this purpose.

Using a feeler gauge, determine and record the clearance at the first valve. If the clearance is incorrect, not being within the range of 0.03 – 0.08 mm (0.001 – 0.003 in), the adjuster pad must be removed and replaced by one of suitable thickness. A special tool is available (Suzuki part No. 09916 – 64510) which may be pushed between the camshaft adjacent to the cam lobe and the raised edge of the cam follower, to allow removal of the shim. If the special tool is not available, a simple substitute may be fabricated from a portion of steel plate.

The final form of the tool which has a handle about 6 inches long, can be seen in the accompanying photograph.

The Suzuki tool may be pushed into position, depressing the cam follower and securing it in a depressed position, in one operation. Where a home-made tool is used, the cam follower may be depressed using a suitable lever placed between the adjuster pad and the cam lobe. The tool may then be inserted to secure the cam follower whilst the adjuster pad is removed. Before installing either type of tool, rotate the cam follower so that the slot in the raised edge is not obscured by the camshaft. Insert a small screwdriver through the slot to displace the adjuster shim.

Adjustment pads are available in 20 sizes ranging from 2.15 mm to 3.10 mm, in increments of 0.05 mm. Each pad is identified by a three digit number etched on the reverse face. The number (eg. 235) indicates that the pad thickness is 2.35 mm thick. To select the correct pad refer to the accompanying table and the example given to the right of that table.

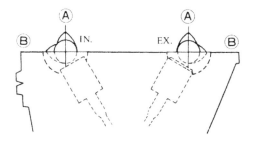

Alternative cam position when checking valve clearances

P/N SUFFIX- Tappet Clearance (mm)	45000	45001	45002	45003	45004	45005	45006	45007	45008	45009	45010	45011	45012	45013	45014	45015	45016	45017	45018	45019
PRESENT SHIM SIZE — mm	2.15	2.20	2.25	2.30	2.35	2.40	2.45	2.50	2.55	2.60	2.65	2.70	2.75	2.80	2.85	2.90	2.95	3.00	3.05	3.10
0.00~0.02		2.15	2.20	2.25	2.30	2.35	2.40	2.45	2.50	2.55	2.60	2.65	2.70	2.75	2.80	2.85	2.90	2.95	3.00	3.05
0.03~0.08							SPECIFIED CLEARANCE/NO. ADJUSTMENT REQUIRED													
0.09~0.13	2.20	2.25	2.30	2.35	2.40	2.45	2.50	2.55	2.60	2.65	2.70	2.75	2.80	2.85	2.90	2.95	3.00	3.05	3.10	
0.14~0.18	2.25	2.30	2.35	2.40	2.45	2.50	2.55	2.60	2.65	2.70	2.75	2.80	2.85	2.90	2.95	3.00	3.05	3.10		
0.19~0.23	2.30	2.35	2.40	2.45	2.50	2.55	2.60	2.65	2.70	2.75	2.80	2.85	2.90	2.95	3.00	3.05	3.10			
0.24~0.28	2.35	2.40	2.45	2.50	2.55	2.60	2.65	2.70	2.75	2.80	2.85	2.90	2.95	3.00	3.05	3.10				
0.29~0.33	2.40	2.45	2.50	2.55	2.60	2.65	2.70	2.75	2.80	2.85	2.90	2.95	3.00	3.05	3.10					
0.34~0.38	2.45	2.50	2.55	2.60	2.65	2.70	2.75	2.80	2.85	2.90	2.95	3.00	3.05	3.10						
0.39~0.43	2.50	2.55	2.60	2.65	2.70	2.75	2.80	2.85	2.90	2.95	3.00	3.05	3.10							
0.44~0.48	2.55	2.60	2.65	2.70	2.75	2.80	2.85	2.90	2.95	3.00	3.05	3.10								
0.49~0.53	2.60	2.65	2.70	2.75	2.80	2.85	2.90	2.95	3.00	3.05	3.10									
0.54~0.58	2.65	2.70	2.75	2.80	2.85	**2.90**	2.95	3.00	3.05	3.10										
0.59~0.63	2.70	2.75	2.80	2.85	2.90	2.95	3.00	3.05	3.10											
0.64~0.68	2.75	2.80	2.85	2.90	2.95	3.00	3.05	3.10												
0.69~0.73	2.80	2.85	2.90	2.95	3.00	3.05	3.10													
0.74~0.78	2.85	2.90	2.95	3.00	3.05	3.10														
0.79~0.83	2.90	2.95	3.00	3.05	3.10															
0.84~0.88	2.95	3.00	3.05	3.10																
0.89~0.93	3.00	3.05	3.10																	
0.94~0.98	3.05	3.10																		
0.99~1.03	3.10																				

I.   Measure tappet clearance. "ENGINE COLD"
II. Measure present shim size.
III. Match clearance in vertical column with present shim size in horizontal column.

EXAMPLE

Tappet clearance is — 0.55 mm
Present shim size — 2.40 mm
Shim size to be used — 2.90 mm

Valve adjustment shim selection table

Although the adjuster pads are available as a complete set their price is prohibitive. It is suggested that pads are purchased individually, after an accurate assessment of requirement has been made. It is possible that some Suzuki service agents will be prepared to exchange needed pads for others of the correct size, provided that the original pads are not worn.

Before installing a replacement pad, lubricate both sides thoroughly with engine oil. Always fit the pad with the identification number downwards, so that it does not become obliterated by the action of the cam. After fitting new adjuster pads, rotate the engine forwards a number of times and then recheck the clearances to verify that no errors have occurred.

Before refitting the cam cover, together with a new gasket, lubricate the camshafts with copious quantities of clean engine oil.

5   Carburettors: adjustment and synchronisation

In order that the engine maintains the best possible performance at all times, the carburettors must always remain correctly adjusted and synchronised. This check is essential and should be carried out at the specified intervals.

Synchronisation and adjustment of the carburettors requires the use of a set of four vacuum gauges or indicators, together with the appropriate adaptors which screw into the inlet tracts and to which are attached the vacuum take-off pipes. The adjustment of the carburettors is critical if smooth running and optimum fuel economy is to be expected and if damage to the engine due to incorrect mixture is to be avoided. Because of this and the prohibitive cost of the gauges required, it is recommended that the machine be returned to a Suzuki Service Agent, who will be able to carry out the work. If the vacuum gauges are available and some previous experience has been gained in their use, refer to Chapter 2 Section 6 for the prescribed adjustment procedure.

Due to the construction of the carburettors fitted to the GS1000 range, the possible adjustments are limited to two items; idling speed adjustment, and throttle cable(s) play. The fuel and air adjustments are pre-set under factory conditions, and as such are not possible items for adjustment under any others circumstances. See Chapter 2, Section 5, paragraph 9 for a more detailed explanation of the factory-set carburettor adjustments.

The idling speed of the engine may be altered by turning the remote adjuster placed below the central two, of the bank of four carburettors. It is recognisable by the large nylon head that it is fitted with. The correct idle speed is 1000 rpm ± 100 rpm. Always adjust the idling speed with the engine having been run so that it is operating at its normal working temperature.

For instruction on adjusting the throttle cables, refer to Chapter 2, Section 6.3. There are two such cables fitted, one to open and one to close the throttle.

6   Cleaning and checking sparking plugs

Remove the sparking plugs and remove the carbon deposits using a wire brush. Clean the electrodes using fine emery paper or cloth, and then reset the gaps to 0.6 – 0.8 mm (0.024 – 0.031 in), using a feeler gauge. Whenever the plugs are removed, make a visual inspection of the condition of the plugs, especially the colour of the carbon deposits. See the sparking plug colour condition chart in Chapter 3. If the standard plug (NGK B8ES or Nippon Denso W24ES) appears deficient; either appearing to be too hot (too soft a grade) or too cold (too hard a grade), then alternative plug types should be installed. If the plugs are apt to run too hot, that is, on examination the electrode is white in colour and has a blistered appearance, a colder range plug is necessary. A set of NGK B9ES or Nippon Denso W27ES plugs should be installed to correct this situation. Conversely, a black, sooty appearance to the plug, may indicate it is running too cold. In this instance a harder grade set of plugs; NGK B7ES or Nippon Denso W22ES, should be fitted to alleviate the problem.

Before replacing the plugs, smear the threads with a small amount of graphite grease to aid future removal.

7   Clutch adjustment

Accurate adjustment of the clutch is necessary to ensure efficient operation of the whole unit. With the type of clutch fitted to the GS1000 range, no adjustment is available on the clutch itself. Adjustment is possible at two points on the clutch cable, one at the operating lever, and a second at the actuating lever on the top of the right-hand side crankcase cover.

Pull back the rubber cover on the handlebar adjuster nut and lock nut. Slacken the locknut, and screw the adjuster fully inwards, to give maximum free play at the lever. Loosen the locknut on the actuating lever on the crankcase cover, and turn the chromed adjuster nut. This tensions the inner cable and reduces the play at the handlebar lever. When the play at the handlebar lever is between 2 – 3 mm, the adjustment is correct. Retighten the locknut at the lower cable and adjuster. If correct adjustment cannot be made by this method, further adjustment, at the handlebar, must be made. Screw out the adjuster until the free play is correct. Secure the locknut, and replace the rubber shroud.

8   Control cable lubrication

Use motor oil or an all purpose oil to lubricate the control cables. A good method for lubricating the cables is shown in the accompanying illustration, using a plasticine funnel. This method has a disadvantage in that the cables usually need removing from the machine. An hydraulic cable oiler which pressurises the lubricant overcomes this problem. Nylon lined cables should not be lubricated; in some cases the oil will cause the lining to swell leading to seizure.

Check valve clearance using a feeler gauge

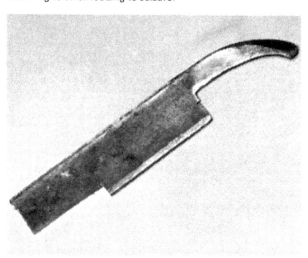

Home-made camfollower depression tool

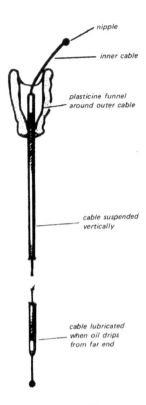

nipple

inner cable

plasticine funnel around outer cable

cable suspended vertically

cable lubricated when oil drips from far end

Control cable oiling

9   Spoke tension – wire wheel models

Check the spokes for tension, by gently tapping each one with a metal object. A loose spoke is identifiable by the low pitch noise emitted when struck. If any one spoke needs considerable tightening, it will be necessary to remove the tyre and inner tube in order to file down the protruding spoke end. This will prevent it from chafing through the rim band and piercing the inner tube.

10   Cylinder head nut and bolt torque check

This check is easiest to carry out after the tappet clearances have been checked and/or adjusted, whilst the petrol tank is removed. Using the tightening sequence given in Chapter 1, Section 44, tighten first the twelve 10 mm nuts to a torque figure of 3.7 kgf m (26 lbf ft), and then the two 6 mm bolts to a torque setting of 0.9 – 1.4 kgf m (6.5 – 10.0 lbf ft).

At this time, it is useful to check the tightness of the exhaust pipe flange bolts. They should be set to a torque figure of 0.9 – 1.4 kgf m (6.5 – 10.0 lbf ft).

11   Steering head bearing check

Make a general check on the condition of the steering head bearings if necessary adjusting them as follows.

Place the machine on the centre stand so that the front wheel is clear of the ground. If necessary apply a weight to the rear portion of the dualseat to prevent the machine tipping forwards. Grasp the front forks near the wheel spindle and push and pull firmly in a fore and aft direction. If play is evident between the upper and lower steering yokes and the head lug casting, the steering head bearings are in need of adjustment. Imprecise handling or a tendency for the front forks to judder may be caused by this fault.

To adjust the bearings, loosen the pinch bolt that passes through the rear of the upper yoke. Immediately below the the upper yoke, on the steering stem, is the pegged adjuster nut. Using a C-spanner, tighten the adjuster nut a little at a time until all play is taken up. **Do Not** overtighten the nut. It is possible to place a pressure of several tons on the head bearings by overtightening, even though the handlebars may seem to turn quite freely. Overtight bearings will cause the machine to roll at low speeds and give imprecise steering. Adjustment is correct if there is no play in the bearings and the handlebars swing to full lock either side when the machine is on the centre stand with the front wheel off the ground. Only a light tap on each end should cause the handlebars to swing.

Eight monthly or every 8000 miles

Again complete the checks listed under the previous routine maintenance interval headings. The following additional tasks are now necessary:
1   Replace the spark plugs. Although in general sparking plugs will continue to function after this mileage, their efficiency will have been reduced. The correct standard plug type is NGK B8ES or Nippon Denso W24 ES. Before fitting, set the gaps to 0.6 – 0.8 mm (0.024 – 0.031 in).
2   Renew the air cleaner element. See Chapter 2, Section 9, paragraphs 1 and 2.
3   Check, and if necessary, adjust the front fork air pressure. See Chapter 4, Section 6, paragraphs 6 – 8.
4   When renewing the oil at this service interval, the sump should be removed and the oil pick-up strainer screen detached for cleaning. The screen is secured to the pick-up chamber by three screws. Wash the screen in clean petrol, using a soft brush.

Yearly or every 12000 miles

Again carry out the tasks and checks listed under the previous routine maintenance interval headings. The following additional checks are now necessary.
1   Check the condition of the contact breaker points assemblies, and renew them if necessary. See Chapter 3, Section 5.
2   Remove and clean the wheel bearings. Renew worn bearings. See Chapter 5, Sections 8 and 14.
3   In addition to the above operations, the various frame and engine fittings should be checked for tightness and lubricated where necessary.
4   In addition to all the foregoing, Suzuki recommend that at two yearly (or 24000 mile) intervals the brake fluid hoses and the fuel lines be renewed.

General adjustment

It may have been noted that no reference has been made in any of the foregoing routine maintenance schedules to brake pad wear or inspection.

Brake pad wear depends largely on the conditions in which the machine is ridden and at what speed. It is difficult therefore to give precise inspection intervals, but it follows that pad wear should be checked more frequently on a hard ridden machine.

The condition of each pad can be checked easily whilst still in situ on the machine. The pads have a red groove around their outer periphery which can be seen from the front of the caliper unit (front brakes), or from the top of the caliper after the inspection cap has been prised off.

Removal and replacement of the brake pads in both the front and rear calipers may be accomplished without removing the respective wheel.

If wear has reduced either or both pads in one caliper down to the red line, the pads should be renewed as a pair. In practice, where a double front disc set-up is used, if one set of pads requires renewal, it will be necessary to renew the other pair, too.

To gain access to the front brake pads for renewal, each caliper unit must be detached from the fork leg, although separation of the caliper from the hydraulic hose is not required. Remove the two bolts which pass through the fork leg into the caliper support bracket and lift the complete caliper unit upwards, off the disc. Remove the single screw and the convolute backing plate from the inner side of the caliper body.

The inner pad is now free and may be displaced towards the centre of the caliper, and lifted out. The outer pad, which abuts against the caliper piston, is not retained positively and may be lifted out. Refit the new pads after coating lightly the periphery of each pad with disc brake assembly grease (silicon grease). Use the grease sparingly, ensuring that it **does not** come in contact with the friction surface of the pad. If necessary, push the piston inwards to increase the clearance between the pads and allow the pads to fit over the disc. To remove the brake pads from the rear brake caliper, pull out the stop pin which passes through the outer end of each pad mounting pin. Displace one mounting pin and remove the two hair springs. Push out the final pin and lift each pad out individually, removing the outer pad first.

Install new pads by reversing the dismantling procedure. The shim fitted to the piston side of each pad must be positioned with the punched arrowmark pointing in the direction of normal wheel travel. If required, push back each piston to give the necessary clearance between each piston and the disc face into which the pad can be inserted.

Quick glance
maintenance adjustments and capacities

Engine oil capacity
    Dry   4.2 lit (8.8/7.4 US/Imp pints)
    Oil change   3.4 lit (7.2/6.0 US/Imp pints)
    Oil and filter change   3.8 lit (8.0/6.6 US/Imp pints)
Front forks USA models, except S   240 cc (8.15/6.76 US/Imp fl oz) per leg
    UK models and S [USA]   260 cc (8.75/7.32 US/Imp fl oz) per leg
Contact breaker gap   0.3 – 0.4 mm (0.012 – 0.016 in)
Spark plug gap   0.6 – 0.8 mm (0.023 – 0.031 in)
Valve clearances (cold)
    Inlet   0.03 – 0.08 mm (0.001 – 0.003 in)
    Exhaust   0.03 – 0.08 mm (0.001 – 0.003 in)
Tyre pressures

	Solo	Pillion
Front	25 psi (1.75 kg cm^2)	28 psi (2.00 kg cm^2)
Rear	28 psi (2.00 kg cm^2)	32 psi (2.25 kg cm^2)

For continuous high-speed riding, the pressures should be increased:

	Solo	Pillion
Front	28 psi (2.00 kg cm^2)	32 psi (2.25 kg cm^2)
Rear	36 psi (2.50 kg cm^2)	40 psi (2.80 kg cm^2)

Recommended lubricants

Component	Lubricant
Engine and transmission	SAE 10W/40 engine oil
Front forks	SAE 10W/20 fork oil
Wheel bearings	Lithium base high-melting point grease
Disc brakes	Hydraulic brake fluid conforming to DOT 3 (US) or SAE J1703 (UK)

Working conditions and tools

When a major overhaul is contemplated, it is important that a clean, well-lit working space is available, equipped with a workbench and vice, and with space for laying out or storing the dismantled assemblies in an orderly manner where they are unlikely to be disturbed. The use of a good workshop will give the satisfaction of work done in comfort and without haste, where there is little chance of the machine being dismantled and reassembled in anything other than clean surroundings. Unfortunately, these ideal working conditions are not always practicable and under these latter circumstances when improvisation is called for, extra care and time will be needed.

The other essential requirement is a comprehensive set of good quality tools. Quality is of prime importance since cheap tools will prove expensive in the long run if they slip or break when in use, causing personal injury or expensive damage to the component being worked on. A good quality tool will last a long time, and more than justify the cost.

For practically all tools, a tool factor is the best source since he will have a very comprehensive range compared with the average garage or accessory shop. Having said that, accessory shops often offer excellent quality tools at discount prices, so it pays to shop around. There are plenty of tools around at reasonable prices, but always aim to purchase items which meet the relevant national safety standards. If in doubt, seek the advice of the shop proprietor or manager before making a purchase.

The basis of any tool kit is a set of open-ended spanners, which can be used on almost any part of the machine to which there is reasonable access. A set of ring spanners makes a useful addition, since they can be used on nuts that are very tight or where access is restricted. Where the cost has to be kept within reasonable bounds, a compromise can be effected with a set of combination spanners – open-ended at one end and having a ring of the same size on the other end. Socket spanners may also be considered a good investment, a basic $3/8$ in or $1/2$ in drive kit comprising a ratchet handle and a small number of socket heads, if money is limited. Additional sockets can be purchased, as and when they are required. Provided they are slim in profile, sockets will reach nuts or bolts that are deeply recessed. When purchasing spanners of any kind, make sure the correct size standard is purchased. Almost all machines manufactured outside the UK and the USA have metric nuts and bolts, whilst those produced in Britain have BSF or BSW sizes. The standard used in USA is AF, which is also found on some of the later British machines. Others tools that should be included in the kit are a range of crosshead screwdrivers, a pair of pliers and a hammer.

When considering the purchase of tools, it should be remembered that by carrying out the work oneself, a large proportion of the normal repair cost, made up by labour charges, will be saved. The economy made on even a minor overhaul will go a long way towards the improvement of a toolkit.

In addition to the basic tool kit, certain additional tools can prove invaluable when they are close to hand, to help speed up a multitude of repetitive jobs. For example, an impact screwdriver will ease the removal of screws that have been tightened by a similar tool, during assembly, without a risk of damaging the screw heads. And, of course, it can be used again to retighten the screws, to ensure an oil or airtight seal results. Circlip pliers have their uses too, since gear pinions, shafts and similar components are frequently retained by circlips that are not too easily displaced by a screwdriver. There are two types of circlip pliers, one for internal and one for external circlips. They may also have straight or right-angled jaws.

One of the most useful of all tools is the torque wrench, a form of spanner that can be adjusted to slip when a measured amount of force is applied to any bolt or nut. Torque wrench settings are given in almost every modern workshop or service manual, where the extent to which a complex component, such as a cylinder head, can be tightened without fear of distortion or leakage. The tightening of bearing caps is yet another example. Overtightening will stretch or even break bolts, necessitating extra work to extract the broken portions.

As may be expected, the more sophisticated the machine, the greater is the number of tools likely to be required if it is to be kept in first class condition by the home mechanic. Unfortunately there are certain jobs which cannot be accomplished successfully without the correct equipment and although there is invariably a specialist who will undertake the work for a fee, the home mechanic will have to dig more deeply in his pocket for the purchase of similar equipment if he does not wish to employ the services of others. Here a word of caution is necessary, since some of these jobs are best left to the expert. Although an electrical multimeter of the AVO type will prove helpful in tracing electrical faults, in inexperienced hands it may irrevocably damage some of the electrical components if a test current is passed through them in the wrong direction. This can apply to the synchronisation of twin or multiple carburettors too, where a certain amount of expertise is needed when setting them up with vacuum gauges. These are, however, exceptions. Some instruments, such as a strobe lamp, are virtually essential when checking the timing of a machine powered by CDI ignition system. In short, do not purchase any of these special items unless you have the experience to use them correctly.

Although this manual shows how components can be removed and replaced without the use of special service tools (unless absolutely essential), it is worthwhile giving consideration to the purchase of the more commonly used tools if the machine is regarded as a long term purchase Whilst the alternative methods suggested will remove and replace parts without risk of damage, the use of the special tools recommended and sold by the manufacturer will invariably save time.

Chapter 1 Engine, clutch and gearbox

Contents

Specifications

Engine

Type	Four cylinder, double overhead camshaft, air-cooled, four-stroke
Bore	70.0 mm (2.756 in)
Stroke	64.8 mm (2.551 in)
Capacity	997 cc (60.8 cu in)
Compression ratio	9.2 : 1
bhp	87 @ 8000 rpm (HC, C, EC, N and L)
	90 @ 8500 rpm (EN and S)

Cylinder head

Max cylinder head warpage . . . .   0.25 mm (0.009 in)
  Valve stem outside diameter:
    Inlet . . . .   6.957 – 6.975 mm (0.273 – 0.274 in)
    Service limit . . . .   6.90 mm (0.271 in)
    Exhaust . . . .   6.944 – 6.960 mm (0.273 – 0.274 in)
    Service limit . . . .   6.80 mm (0.267 in)
  Valve head diameter:
    Inlet . . . .   38 mm (1.496 in)
    Exhaust . . . .   32 mm (1.259 in)
  Valve guide internal diameter:
    Inlet/exhaust . . . .   7.000 – 7.015 mm (0.275 – 0.276 in)
    Service limit; inlet . . . .   7.09 mm (0.2791 in)
    Service limit: exhaust . . . .   7.10 mm (0.2795 in)
  Valve stem to guide clearance:
    Inlet . . . .   0.025 – 0.058 mm (0.0009 to 0.0022 in)
    Service limit . . . .   0.09 mm (0.0035 in)
    Exhaust . . . .   0.040 – 0.071 mm (0.0015 – 0.0027 in)
    Service limit . . . .   0.10mm (0.0039 in)
  Valve spring minimum free length
    Inner . . . .   33.8 mm (1.330 in)
    Outer . . . .   41.5 mm (1.633 in)

Camshafts

  Inlet:
    Standard lift . . . .   8 mm (0.314 in)
    Overall lobe height . . . .   36.32 – 36.36 mm (1.429 – 1.431 in)
    Service limit . . . .   36.00 mm (1.417 in)
    Journal inside diameter . . . .   21.959 – 21.980 mm (0.864 – 0.865 in)
    Service limit . . . .   21.920 mm (0.862 in)
    Journal outside diameter . . . .   22.000 – 22.013 mm (0.8661 – 0.8666 in)
    Service limit . . . .   22.100 mm (0.870 in)
  Exhaust:
    Standard lift . . . .   7.5 mm (0.295 in)
    Overall lobe height . . . .   35.77 – 35.81 mm (1.408 – 1.409 in)
    Service limit . . . .   35.50 mm (1.397 in)
    Journal inside diameter . . . .   21.959 – 21.980 mm (0.864 – 0.865 in)
    Service limit . . . .   21.920 mm (0.862)
    Journal outside diameter . . . .   22.000 – 22.013 mm (0.8661 – 0.8666 in)
    Service limit . . . .   22.100 mm (0.870 in)
  Camshaft/journal clearance:
    Standard . . . .   0.020 – 0.054 mm (0.0008 – 0.0021 in)
    Service limit . . . .   0.150 mm (0.0059 in)
  Camshaft run-out:
    Standard . . . .   0.03 mm (0.0012 in)
    Service limit . . . .   0.01 mm (0.0039 in)
  Camshaft chain size DID219FTS
  No of chain links   120

Piston and rings

Piston diameter (standard) . . . .   69.945 – 69.960 mm (2.753 – 2.754 in)
  Service limit . . . .   69.800 mm (2.748 in)
  Gudgeon pin OD . . . .   17.995 – 18.000 mm (0.7084 – 0.7086 in)
  Service limit . . . .   17.96 mm (0.707 in)
  Piston ring free end gap:
  Top and 2nd ring . . . .   8.5 mm (0.33 in)
  Service limit . . . .   6.5 mm (0.25 in)
  Piston ring end gap:
  Top and 2nd ring . . . .   0.15 – 0.35 mm (0.005 – 0.013 in)
  Service limit . . . .   0.65 mm (0.025 in)
  Piston ring thickness:
  Top . . . .   1.175 – 1.190 (0.0463 – 0.0469 in)
  Service limit . . . .   1.10 mm (0.043 in)
    2nd . . . .   1.170 – 1.190 mm (0.0460 – 0.0469 in)
    Service limit . . . .   1.10 mm (0.043 in)
  Piston ring side clearance:
    Top . . . .   0.020 – 0.055 mm (0.0008 – 0.0022 in)
    Service limit . . . .   0.18 mm (0.007 in)
    2nd . . . .   0.020 – 0.060 mm (0.0008 – 0.0024 in)
    Service limit . . . .   0.18 mm (0.007 in)
    Oil control ring (max) . . . .   0.15 mm (0.005 in)

Cylinder barrels

Bore diameter	70.000 – 70.015 mm (2.755 – 2.756 in)
Service limit	70.100 mm (2.759 in)
Cylinder/piston clearance	0.050 – 0.060 mm (0.0020 – 0.0024 in)
Compression pressure	9 – 13 kg/cm^2 (128 – 184 psi)
Service limit	7 kg/cm^2 (100 psi)

Crankshaft and connecting rods

Crankshaft run-out	0 – 0.07 mm (0 – 0.002 in)
Service limit	0.10 mm (0.003 in)
Connecting rod deflection (max)	3.0 mm (0.11 in)
Connecting rod axial float	0.10 – 0.65 mm (0.003 – 0.025 in)
Service limit	1.0 mm (0.039 in)
Small-end bore	18.006 – 18.014 mm (0.708 – 0.709 in)
Service limit	18.07 mm (0.71 in)

Clutch

Type	Wet, multi-plate
No of plates	
Plain	7
Friction	8
Plain plate thickness (min)	1.6 mm (0.06 in)
Friction plate thickness	2.7 – 2.9 mm (0.10 – 0.11 in)
Service limit	2.5 mm (0.009 in)
No of springs	6
Spring free length	39.0 – 40.5 mm (1.535 – 1.590 in)
Service limit	38.5 mm (1.51 in)

Gearbox

Type	5-speed, constant mesh
Gear ratios: overall:	
1st gear	2.500 : 1
2nd gear	1.777 : 1
3rd gear	1.380 : 1
4th gear	1.125 : 1
5th gear	0.961 : 1
Primary reduction	1.775 : 1
Final drive ratio	2.800 : 1 (15/42)

Main torque wrench settings

Cylinder head:	
6 mm bolts	0.9 kgf m (6.5 lbf ft)
10 mm nuts	3.7 kgf m (26 lbf ft)
Cylinder head cover bolts (6 mm)	0.9 kgf m (6.5 lbf ft)
Engine mounting bolts:	
8 mm bolts	2.5 kgf m (18 lbf ft)
10 mm bolts	3.5 kgf m (25 lbf ft)
Crankcase bolts:	
6 mm bolts	1.0 kgf m (7 lbf ft)
8 mm bolts	2.0 kgf m (14 lbf ft)
Camshaft cap nuts (6 mm)	1.0 kgf m (7 lbf ft)
Clutch spring bolts (6 mm)	1.1 – 1.3 kgf m (8 – 9 lbf ft)
Clutch centre nut (24 mm)	5.0 – 7.0 kgf m (36 – 51 lbf ft)
Alternator rotor centre bolt (12 mm)	9.0 – 10.0 kgf m (65 – 72 lbf ft)
Engine sprocket nut	9.0 – 10.0 kgf m (65 – 72 lbf ft)

1 General description

The engine unit fitted to the Suzuki GS1000 range is of the four-cylinder, air cooled, in-line type, fitted transversely across the frame. The valves are operated by double overhead camshafts driven from the crankshaft by a centre chain. The two camshafts are located in the cylinder head casting and the camshaft chain drive operates through a cast-in tunnel between the two innermost cylinders. Adjustment of the chain is effected by a chain tensioner, fitted to the rear of the cylinder block. The chain tensioner is of the automatic, self-adjusting type, maintaining correct tension on the chain, compensating for chain wear, after the initial adjustment has been made during assembly.

The engine/gear unit is of aluminium alloy construction, with the crankcases dividing horizontally.

The lubrication is of the pressure feed, wet-sump type. The system incorporates a gear driven oil pump, an oil filter, a safety by-pass valve, and an oil pressure switch. Oil vapours created in the crankcase are vented through an oil breather to the air cleaner case, where they are recirculated into the crankcase providing an oil tight system.

An Eaton trochoid oil pump is fitted, driven by a gear pinion to the rear of the clutch outer drum. Oil is picked up from the sump via a chamber integral with the upper crankcase half, the mouth of which is closed by a detachable wire mesh screen. The screen protects the oil pump from any larger impurities which may have contaminated the oil. The oil pump forces the oil, under pressure, through a full flow paper-element oil filter which is housed within a chamber at the front of the crankcase. In the event of the filter becoming blocked, a by-pass valve is included which prevents cessation of the oil flow by opening at a preset pressure.

After passing through the oil filter, the oil flow is separated into three branch systems. The main system feeds the crankshaft main bearings and the big end bearings, and the two remaining systems supply the camshafts and valve and the gearbox shafts and pinions. Returning oil from the engine and gearbox components falls under gravity to the sump, where it is picked up once more by the oil pump and the cycle is repeated.

2 Operations with the engine/gearbox unit in the frame

1 It is not necessary to remove the engine from the frame to carry out certain operations; in fact it can be an advantage. Tasks that can be carried out with the engine in situ are as follows;
a) Removal of cylinder head, cylinder block and pistons.
b) Removal of the clutch.
c) Removal of the alternator and starter motor.
d) Removal of the carburettors.
e) Removal of the gearchange external selector mechanism.
f) Removal of the oil pump and oil filter.

2 When several tasks have to be undertaken simultaneously, it will probably be advantageous to remove the complete engine unit from the frame, an operation that should take about an hour and a half. This gives the advantage of much better access and more working space.

3 Operations with the engine/gearbox unit removed from the frame

a) Removal of the crankshaft assembly, complete with main bearings and connecting rod assemblies.
b) Removal of the gearbox components including the gearchange internal selection mechanism.
c) Removal of the kickstart shaft and engagement mechanism.

4 Method of engine/gearbox removal

As described previously, the engine and gearbox are a built in unit and it is necessary to remove the unit complete in order to gain access to either assembly. Separation of the crankcase is achieved after the engine unit has been removed from the frame and refitting cannot take place until the engine/gear unit is assembled completely. Access to the gearbox is not available until the engine has been dismantled and vice-versa in the case of attention to the bottom end of the engine. Fortunately, the task is made easy by arranging the crankcase to separate horizontally.

5 Removing the engine/gearbox unit

1 Place the machine on the centre stand, so that it is standing firmly on level ground. Place a receptacle that will hold at least a gallon under the crankcase and remove the drain plug so that the oil will drain off. It is preferable to do this whilst the engine is warm, so that the oil will drain more readily.

2 Transfer the container so that it rests below the oil filter chamber at the front of the engine. Remove the three domed nuts and the washers and detach the cover. The cover is spring loaded by the filter element retainer spring and so should be released in a controlled manner. Lift out the spring and the element.

3 Raise the dualseat and remove the single retaining bolt from the air cleaner case. The filter element can be left in situ in the case. Disconnect the hose that connects the underseat box to the forward filter case, by releasing the bolt on the hose clip.

4 Detach the left-hand and right-hand side covers. Each is retained at the lower edge by a single screw, and at the upper edge by two projections which locate with two rubber covered hook tabs on the frame.

5 Remove the air filter element box by lifting the box away upwards.

6 The battery is now visible. Disconnect both battery leads from the terminals. Isolating the electrical system in this way will prevent accidental shorting of wires which are subsequently disconnected. If it is expected that the machine is to be unused for a protracted length of time, the battery should be removed at this stage and given a refresher charge from an independent source at approximately monthly intervals. Lift the battery from the carrier after displacing the retaining strap and pulling the small breather pipe from the union at the side of the battery.

7 Place the petrol tap lever to the On or Reserve position and disconnect from the tap unions, the petrol feed pipe and the small vacuum pipe which controls the tap diaphragm. The feed pipe is secured by a spring clip, the ears of which should be squeezed together to release the grip on the pipe. Remove the single bolt which passes through the lug at the rear of the petrol tank and into the frame. The tank is supported at the front by two steel cups which locate with a rubber buffer each side of the frame top tube. Before removing the petrol tank the fuel gauge leads must be disconnected at their snap connectors. They are situated at the front on the left-hand side of the tank. Ease the tank rearwards until the cups clear the buffers and then lift it away. Drainage of the tank is not strictly necessary although to do so will reduce the overall weight and hence facilitate removal. A full tank will weight in the region of 50 lb. On the GS1000 L model, detach the two plastic control cable shrouds, which are fitted to each side of the frame top tube as it leaves the steering stem.

8 Pull off the engine breather hose from the breather cover union on top of the cam cover. The hose is secured by a spring clip. Disconnect both throttle cables from the operating pulley at the carburettors. Each may be detached in a similar manner. Loosen the upper and lower locknuts on the cable adjuster screw and displace the adjuster and outer cable from the abutment bracket. Rotate the pulley until the nipple can be pushed out.

9 Loosen the screw clips which secure the carburettors to the inlet stubs, and those securing the air filter hoses to the carburettor mouths. Remove the two front air filter case retaining bolts and ease the air filter case rearwards, allowing the hoses to leave the carburettors. Because of the limited space in this area, and the large size of the air filter case, some 'wriggling' and careful manoeuvring will probably be necessary to allow the carburettors, and then the filter case, to pull clear. Pull the carburettors from the inlet stubs and remove them as a complete unit. The front air filter case can now be displaced. This case can be removed only AFTER the carburettors have been removed.

10 Detach both forward footrests, each of which is retained by two bolts passing into the frame. From the left-hand side of the machine remove the gearchange lever from its splined operating shaft. The lever is retained by a pinch bolt which must be unscrewed fully before the lever can be displaced. Remove the screws which secure the final drive sprocket cover, and detach the cover. The screws retaining this cover, in common with the screws retaining the other external covers, will undoubtedly be tight if they have not been removed previously. This is due to the method of factory assembly, with the screws being tightened by means of air-assisted tools. This being the case, an impact driver will almost certainly be necessary to avoid total destruction of the crosshead type screws fitted to these Suzukis. Because the final drive chain is of the endless type (ie. has no spring link) the chain and primary drive sprocket must be removed simultaneously from the gearbox output shaft. Separating the chain, by use of a rivet extractor, should not be attempted on this type of chain. The resultant weak point would be a major safety hazard; the chain separating under a high power loading, which this machine is capable of generating, would be both extremely dangerous and very costly in subse-

quent mechanical repairs. Bend down the tab washer which secures the sprocket nut and remove the nut. To prevent rotation of the shaft whilst loosening the nut apply the rear brake fully. The sprocket can now be eased off the splined shaft, still meshed with the chain, and put to one side, for examination later.

11 From the right-hand side of the machine remove the clutch cable from its operating arm on the top of the clutch cover. To disconnect the cable at the operating mechanism arm requires some slack in the cable. Displace the rubber shroud on the handlebar control lever and screw the adjuster screw inwards fully. It is not necessary to disconnect the cable from the handlebar lever unless examination is required. Pull the rubber boot back on the operating arm mechanism and loosen the locknut to allow the adjusting nut to be slackened fully. The cable can now be separated from the release arm, after first displacing the split pin which retains the cable end nipple. The chromed release arm is retained by one bolt. Release the bolt and displace the release arm by raising it upwards.

12 Remove the rear brake pedal from the splined shaft, after unscrewing the pinch bolt fully. Prise the end of the pedal return spring from the anchor peg, with the aid of a screwdriver blade, and remove the pedal and spring together.

13 Slacken the bolts holding the four exhaust pipes to silencer joint clamps. Slide each clamp back up the pipe to clear the joint. Starting with the two outer pipes, each of which are separate, the exhaust system can now be displaced. Removal of the silencers is not strictly necessary to allow the exhaust system to be separated from the engine. It may, however, be easier to detach them, before lifting the system away from the machine, as the weight is considerable. Their removal will also avoid unnecessary damage to what are expensive items. The silencers are each retained by two bolts. Slacken evenly and then remove the two bolts securing each finned exhaust port flange. Slide the two outside flanges down the pipes and then ease the pipes from the exhaust ports and, if they have not been detached, the silencers. The two inner pipes form an integral unit with the balance pipe under the engine. When sliding down the finned flanges of the centre pipes, the split seating collars should be caught as they drop free. These split collars are not fitted on the individual outer pipes. To aid reassembly, the outer pipes are stamped R and L.

14 Remove the cover which encloses the starter motor, and is held by two screws. Detach the heavy cable from the terminal projecting from the starter motor body. Pull the lead from the

top of the oil pressure warning switch. The lead is a push fit. Pull both leads from where they are mounted and lodge them out of harm's way, towards the rear of the machine. Follow the leads from the alternator up to the rubber shroud to the rear of the battery carrier. Peel back the shroud and disconnect the alternator leads at their individual snap connectors. The two leads from the contact breakers are similarly protected by a rubber sleeve, forward of the left-hand rear down tube. Disconnect the black wire and white wire. Follow the lead from the neutral indicator switch located in the gearbox left-hand wall and disconnect the block connector and single lead. All wires leading from the engine should be arranged at the rear of the engine so that they do not become snagged on engine removal.

15 To improve the clearance between the top of the engine and the upper frame tubes, and so aid engine removal, remove the two horns. Disconnect the two snap connectors on each and remove the single retaining bolt on each from the frame tube. For the same reason, that of limited clearance between the engine and frame tubes, the coils can be removed. The two coils, each held by two bolts, should be removed complete with the HT leads attached. Each lead is numbered to aid correct replacement. Disconnect the main wire harness at the large block connector. Detach the tachometer cable from its drive unit at the front of the cylinder head.

16 The engine/gearbox unit is an extremely large, heavy and cumbersome component, and as such is not easily manhandled. It is recommended, therefore, that at least three people are present when lifting the engine from place. This alleviates the risk of damage to both the engine and the operator. Before attempting to remove any of the engine mounting bolts, the aquisition of a jack will be very beneficial. By positioning the jack below the sump, much of the weight can be taken when the mounting bolts are removed. Remove first the front three mounting bolts, complete with their brackets. Then remove the three upper rear mounting bolts and the mounting bracket on the right side only. Remove the single lower rear nut on the right, and the bolt on the left-hand side. Lastly, remove the bolts fitted below the engine on either side, and their brackets. Note that these centrally positioned mounting bolts are fitted with triangular threaded plates in place of normal nuts, which are restrained within recesses in the crankcase casting. After removal of all the bolts, the engine will have settled on to the jack allowing the operator, and his assistants, time to find a handhold. Lift the engine up gradually and out of the frame from the right-hand side.

5.6 Remove the battery from its carrier

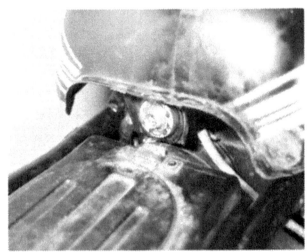

5.7 Remove the single petrol tank retaining bolt

5.8 Disconnect each of the throttle cables

5.9a Remove the two front air filter case retaining bolts

5.9b The front air filter case can only be displaced with the carburettors removed

5.10a Detach the forward footrests; each is held by two bolts

5.10b Remove the gearchange lever from its splined shaft

5.10c Withdraw sprocket still meshed with the chain

5.11 Free the clutch cable from the actuating lever

5.12 Unscrew the pinch bolt that retains the rear brake lever

5.13a Slacken and pull back the exhaust clamps to clear packing

5.13b Remove the silencer retaining bolts at the rear ...

5.13c ... and separate the balance pipe by removing the single bolt

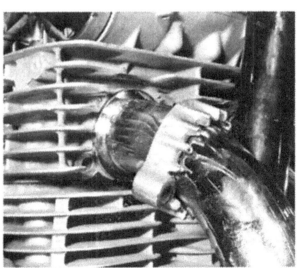

5.13d Remove flange bolts and detach the exhaust pipes

5.15a Remove the coils after removing the two bolts on each

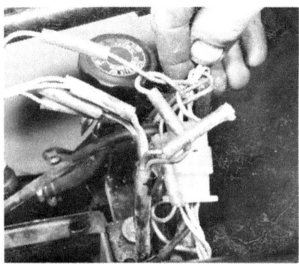

5.15b Disconnect the main wire harness at the large block connector

5.16a Position a jack below the sump

5.16b Remove first the front three mounting bolts, complete with their brackets ...

5.16c ... then the three upper rear bolts, on the right-hand side only

5.16d Remove the single lower rear nut on the right-hand side ...

5.16e ... and the bolt on the left-hand side

5.16f Three bolts and brackets screw into plate nuts

6   Dismantling the engine and gearbox: general

1   Before commencing work on the engine unit, the external surfaces should be cleaned thoroughly. A motor cycle engine has very little protection from road grit and other foreign matter, which will find its way into the dismantled engine if this simple precaution is not taken. One of the proprietary cleaning compounds, such as Gunk or Jizer can be used to good effect, particularly if the compound is permitted to work into the film of oil and grease before it is washed away. Special care is necessary when washing down to prevent water from entering the now exposed parts of the engine unit.
2   Never use undue force to remove any stubborn part unless specific mention is made of this requirement. There is invariably good reason why a part is difficult to remove, often because the dismantling operation has been tackled in the wrong sequence. Dismantling will be made easier if a simple engine stand is constructed to correspond with the engine mounting points. This arrangement will permit the complete unit to be clamped rigidly to the workbench, leaving both hands free.

7   Dismantling the engine unit: removing the camshaft cover and the camshafts

1   Remove the four end caps from the camshaft cover, and unscrew the bolts holding the camshaft cover in position. If the cover is stuck firmly to the gasket, a rawhide mallet may be used to break the seal. Use the mallet judiciously, striking only those parts of the cover which are well strengthened by lugs.
2   Unscrew the sparking plugs and remove the contact breaker cover from the engine right-hand casing. Using a spanner applied to the large hexagon fitted to the end of the contact braker cam, rotate the engine forwards until the piston in the left-hand cylinder (No. 1 cylinder) is at top dead centre. TDC may be found by viewing the timing marks on the Automatic Timing Unit (ATU) through the inspection aperture in the contact breaker stator plate. Turn the engine until the T mark, which is to the left of the F1-4 mark, is in alignment with the index pointer in the casing.
3   Removal of the automatic cam chain tensioner must be carried out in a special sequence. Commence by loosening the locknut securing the grubscrew in the left-hand side of the tensioner body. Turn the screw inwards so that it tightens against the plunger within the body. The tensioner can now be removed without the spring loaded plunger being displaced.
4   Detach the rubber chain guide from its position in the camchain tunnel between the sprockets. This guide will lift out.
5   Having removed the chain guide, the camshafts may be removed individually, without separating the cam chain. It will

be noted that no matter in what position the engine is placed, at least one cam lobe will be depressing one valve and spring to some extent. To prevent uneven stress to the camshafts when removing the camshaft bearing caps. Suzuki recommend that a large self-grip wrench be used to hold the camshaft down. A G-clamp of suitable size will make a substitute if a wrench is not available. Fit the wrench as shown in the accompanying photograph, ensuring that it is so placed that slipping is not possible. Loosen evenly the four bolts holding each of the cams' two bearing caps. Note that each cap is marked A, B, C, or D. The casing below each cap is marked similarly, to enable the caps to be refitted in their original locations and positions. Displace the clamping tool to free the camshafts.
6   If a suitable clamping tool is not available, it is acceptable to slacken the bearing cap bolts without restraining the camshafts, provided extreme caution is exercised. The bolts must be loosened evenly a little at a time, so that neither the camshaft nor the bearing caps are allowed to hit.
7   Lift the cam chain off the sprocket and remove the camshaft, complete with the sprocket. Repeat the procedure for the other camshaft. Removal of the sprockets is not required unless the components require renewal.
8   If a top-end overhaul only is anticipated, the cam chain should not be allowed to fall down into the chain tunnel, as retrieval can be very difficult. Insert a long bar through the chain so that it rests on the cylinder head or use a length of stout wire secured to an adjacent stud or bolt hole.

7.1 Exercise care when separating cam cover from cylinder head

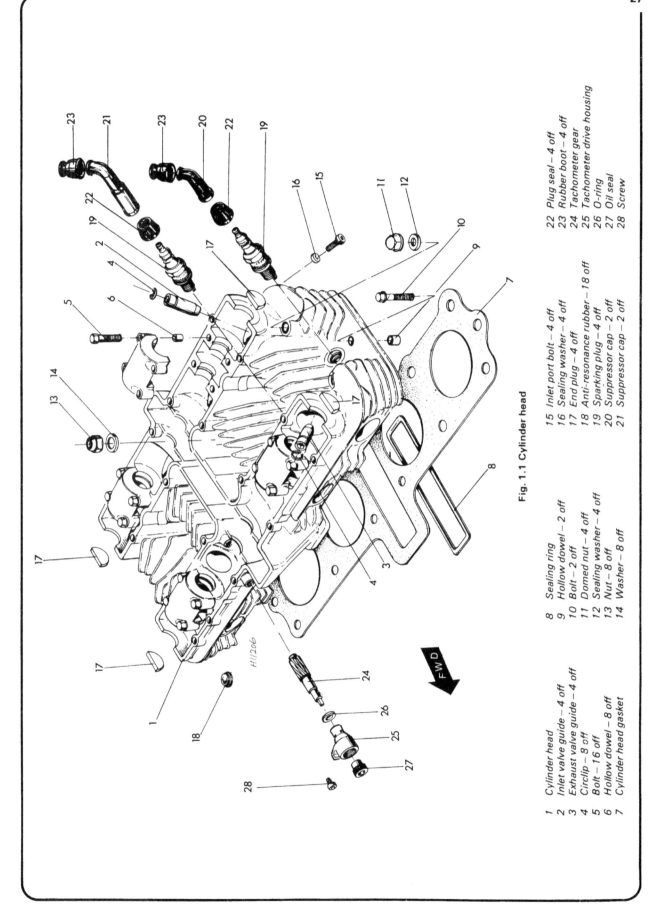

Fig. 1.1 Cylinder head

1 Cylinder head
2 Inlet valve guide – 4 off
3 Exhaust valve guide – 4 off
4 Circlip – 8 off
5 Bolt – 16 off
6 Hollow dowel – 8 off
7 Cylinder head gasket

8 Sealing ring
9 Hollow dowel – 2 off
10 Bolt – 2 off
11 Domed nut – 4 off
12 Sealing washer – 4 off
13 Nut – 8 off
14 Washer – 8 off

15 Inlet port bolt – 4 off
16 Sealing washer – 4 off
17 End plug – 4 off
18 Anti-resonance rubber – 18 off
19 Sparking plug – 4 off
20 Suppressor cap – 2 off
21 Suppressor cap – 2 off

22 Plug seal – 4 off
23 Rubber boot – 4 off
24 Tachometer gear
25 Tachometer drive housing
26 O-ring
27 Oil seal
28 Screw

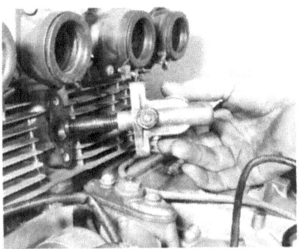
7.3 Remove the chain tensioner unit

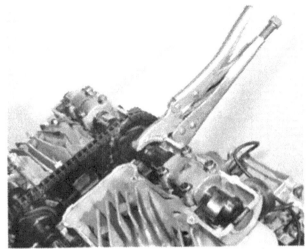
7.5 Suzuki recommend this method of camshaft retention

7.7 Remove the camshaft complete with sprocket

8   Dismantling the engine unit: removing the cylinder head, cylinder block and pistons

1   Do not disturb the cam followers and adjusters at this juncture. These components should be left in place until the examination stage. The cylinder head is retained by twelve 10 mm nuts and two 6 mm bolts, one at each end of the head. Later models have a third bolt fitted at the front of the cam chain tunnel. Slacken off the bolts first and then slacken the twelve nuts in the reverse order of that given in Fig. No 1.15 which accompanies Section 44. To aid correct dismantling the number of the nut is stamped next to it in the head casting. The nuts must be loosened evenly, in this sequence, to avoid stressing the cylinder head cast.

2   Separate the cylinder head from the gasket, using a rawhide mallet. Strike only those portions of the casing which are adequately supported, taking special care not to damage the fins. Under no circumstances should levers be used between the mating surfaces of the cylinder head and cylinder block in an effort to facilitate separation. This action will lead to damage to the faces with subsequent risk of leakage. When lifting the cylinder head from position, the cam chain must be guided through the central tunnel and prevented from falling free. An extra pair of hands is beneficial at this stage.

3   Separate the cylinder block from the base gasket, using the technique described for the cylinder head. Once again, levers should not be used. Lift the cylinder block upwards along the holding down studs until the piston skirts are visible but the piston rings are still obscured by the cylinder bore spigots. If a top-end overhaul only is to be carried out the crankcase mouths must be padded with clean rags to prevent pieces of broken piston ring from falling into the crankcase. The padding will also prevent the ingress of foreign matter during further dismantling or work. With the padding in place, lift the cylinder block off the pistons as squarely as possible to prevent the pistons from tying in the bore. Catch the pistons as they emerge from each bore, to prevent damage occurring.

4   Before removing the pistons, each should be marked using a metal scribe on the inside of the skirt. Number the pistons from 1 to 4, so that if they are to be re-used they may be fitted in their original locations. An arrow mark cast on each piston crown indicates the correct direction in which the piston should be replaced.

5   Remove both circlips from each piston boss and discard them. Circlips should never be re-used if risk of displacement is to be obviated.

6   Using a drift of the correct diameter, tap each gudgeon pin out of the piston bosses until the piston can be lifted off the connecting rod, complete with rings. Make sure the piston is supported during this operation, or there is risk of bending the connecting rod.

7   If the gudgeon pin is a tight fit do not resort to force. Warm the piston by placing a rag soaked in hot water on the crown, so that the piston bosses will expand and release their grip on the pin.

8   Each piston is fitted with three piston rings; two compression and one oil scraper. To remove the rings, spread the ends sufficiently with the thumbs to allow each ring to be eased from its groove and lifted clear of the piston. This is a very delicate operation necessitating great care. Piston rings are very brittle and will break easily.

An alternative method of removing piston rings safely, especially when the rings are gummed up, is shown in the accompanying illustration. Use three narrow strips of tin cut from an old oil can, slipped between the rings and the piston.

9   Unscrew the single bolt which secures the cam chain rear guide blade and lift the blade from position.

8.1a Remove the 6 mm bolts at each end of the cylinder head

8.1b The 10 mm nuts have numbers stamped in the head to aid correct dismantling order

8.2 Keep cylinder head square to studs when removing

8.3 Lift the cylinder block off the pistons

8.5 Prise out the piston circlips and ...

8.6 ... push out the gudgeon pin to free the piston

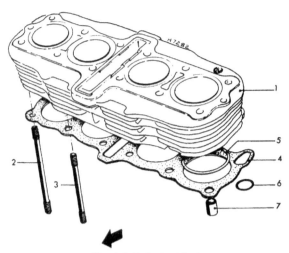

Fig. 1.2 Cylinder block

1 Cylinder block	5 O-ring – 4 off
2 Stud – 4 off	6 O-ring – 2 off
3 Stud – 8 off	7 Dowel pin – 2 off
4 Cylinder base gasket	

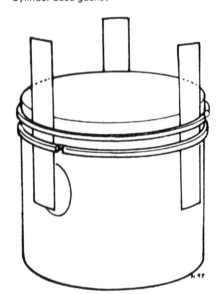

Fig. 1.3 Freeing gummed rings

9 Dismantling the engine unit: removing the contact breakers and automatic timing unit (ATU)

1   Free the contact breaker wiring leads from the clips positioned on the underside of the engine and displace the grommet from the casing wall, after first removing the points cover and gasket. Place a spanner on the engine turning hexagon, and loosen the contact breaker cam centre bolt. Remove the bolt and also the hexagon.

2   Before removing the contact breaker main baseplate, mark the position of the plate relative to the main engine casing. This can be done by putting a scratch mark on the plate, that continues onto a web of the casing (see accompanying photograph), or by using a punch to do the the same thing. This is important because on reassembly the marks can be realigned, so simplifying the ignition timing checking and adjustment operation. Unscrew the three screws which pass through the elongated holes in the main baseplate edge. Detach the contact

breaker assembly as a complete unit and remove the ignition timing index plate which is fitted behind the base plate. Lift the automatic advance/timing unit (ATU) from position, noting the drive pin in the crankshaft end which locates with a notch on the end face of the ATU. Check the fit of the drive pin in the crankshaft, removing it if loose, to avoid accidental loss.

3   At this stage the neutral position indicator switch (or gear position switch) and its lead can be detached. The switch is held by two screws. Care should be exercised not to lose the switch contact and its small spring.

10 Dismantling the engine unit: removing the starter motor and intermediate gear

1   Remove the two screws which pass through the starter motor end cap flange into the crankcase. Ease the starter motor towards the right-hand side of the engine until the motor boss leaves the casing. Lift the starter motor up at the rear and out of the compartment. If the starter motor boss is tight in the casing, insert a wooden lever between the front of the motor and the casing wall. Use the lever to push the motor from place.

2   Free the alternator wiring lead from the guideway in the top of the crankcase. On removal of the alternator cover the lead must be fed through the casing wall because the leads are connected to the alternator stator. Loosen evenly and remove the alternator cover retaining screws. Lift the cover away and pull the lead through. Pass the connectors through the aperture individually.

3   Withdraw the intermediate gear spindle and lift the spindle and gear from position. Note the two shims, one of which is fitted to the spindle, either side of the gear.

11 Dismantling the engine unit: removing the alternator, rotor and starter motor clutch

1   Loosen and remove the alternator rotor centre bolt after applying a spanner to the two flats on the rotor centre boss to prevent rotation. The rotor is a very tight fit on the tapered crankshaft end and will require pulling from position. The rotor boss is threaded internally to take a slide hammer. This tool consists of a headed shaft upon which is placed a free sliding steel weight. After screwing the shaft into the centre boss the steel weight is slid forcibly along the shaft away from the rotor until it contacts the shaft head. The force imparted should release the taper. If the correct slide hammer is not readily available, an alternative tool of the same type may be constructed using the pivot shaft which supports the machine's rear swinging arm. The shaft should be pushed out, after removal of the nut, using a suitable substitute rod of approximately the correct diameter. A large socket spanner makes an ideal sliding weight, though in the absence of this an alternative must be found. If the rotor appears not to be yielding at all to this method of removal, a further method can be employed. Insert a stout screw or bolt, plus two or three washers into the centre boss. The screw, or bolt, will then bear onto the end of the crankshaft. Insert the swinging arm pivot as described and proceed to operate the sliding hammer technique.

2   It may be found that the alternator rotor resists all attempts at removal. If this is the case, expert advice should be sought because the pressed up crankshaft and the rotor itself may suffer severe damage if excess force is employed. Provided that the rotor does not require attention and that the starter clutch and crankshaft oil seal are in good condition, these two components may be left on the crankshaft as their presence does not obstruct further dismantling. If work on the crankshaft and bearings is envisaged, the crankshaft will in any event require returning to a Suzuki Service Agent who may be entrusted to remove the alternator at the same time.

3   To continue dismantling, pull the rotor off the shaft complete with the starter clutch to which it is attached. Slide the starter motor clutch gear off the shaft and remove the thrust washer.

9.2a Scratch mark contact breaker stator plate and crankcase web before ...

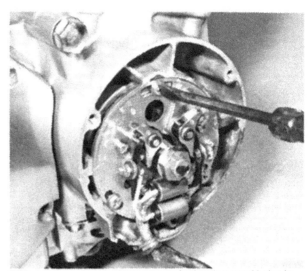

9.2b ... unscrewing the three screws in the elongated holes in the main baseplate edge

9.2c With the contact breaker assembly removed, detach the timing index plate fitted behind

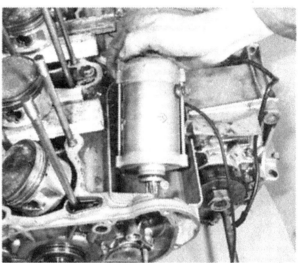

10.1 Lift starter motor out very carefully

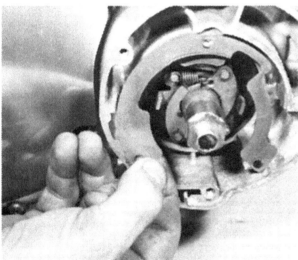

10.2a Lift off alternator cover and ...

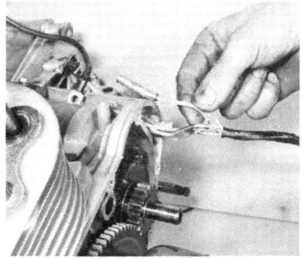

10.2b ... feed the wires through the casing wall

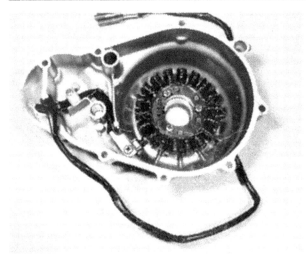

10.2c Alternator cover and stator, general view

10.3 Withdraw spindle and remove intermediate gear

11.1 Improvised slide hammer made from swinging arm pivot and a large socket

11.3 Pull off the rotor and starter clutch as a unit

12 Dismantling the engine unit: removing the clutch

1   Loosen evenly and remove the *eleven* screws which secure the primary drive cover to the right-hand side of the engine. At the same time remove the gasket on the cover.

2   Unscrew the clutch pressure spring bolts evenly and remove them, together with the springs. Remove the clutch pressure plate and then withdraw the clutch plates one at a time, noting the alternating sequence. Pull the clutch operating thrust piece from the centre of the hollow shaft and then displace the pushrod, scroll end first. Note the thrust bearing and washer which fit on the thrust piece. These components are easily mislaid.

3   Bend down the ear of the tab washer which secures the clutch centre nut. The nut may be very tight, and in order to prevent the shaft rotating adopt the following procedure. Prevent the clutch centre boss from rotating by locking it in position with a length of steel plate engaged with one of the teeth on the clutch drum, and the inside of the clutch outer casing. To prevent damage to the outer casing a thin length of wood should be positioned between the steel plate and the casing (see accompanying photograph). Care must be taken, when carrying out this operation, that the steel plate does not slip off the clutch boss tooth that it is secured against. With the nut slackened, it can now be removed, followed by the tab washer.

4   Withdraw the clutch centre boss and remove the thrust washer from the shaft. The clutch outer drum revolves on a caged needle roller bearing, supported on a large central spacer. To enable the edge of the outer drum to clear the casing on removal, the spacer must be withdrawn. Use a clutch spring bolt screwed into one of the threaded holes in the spacer as a means of removal. Pull out the needle bearing and then lift the outer drum towards the rear and out of the primary drive case. The oil pump drive gear fitted to the rear of the clutch may now be removed together with the bearing, spacer and backing washer.

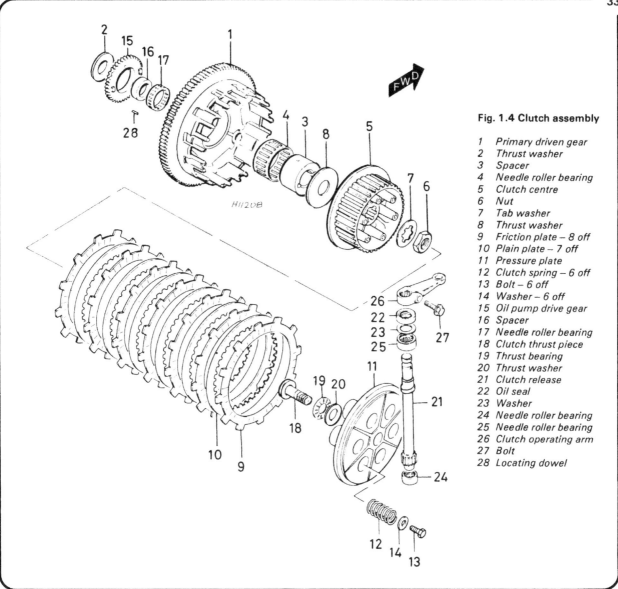

H11208

Fig. 1.4 Clutch assembly

1   Primary driven gear
2   Thrust washer
3   Spacer
4   Needle roller bearing
5   Clutch centre
6   Nut
7   Tab washer
8   Thrust washer
9   Friction plate – 8 off
10 Plain plate – 7 off
11 Pressure plate
12 Clutch spring – 6 off
13 Bolt – 6 off
14 Washer – 6 off
15 Oil pump drive gear
16 Spacer
17 Needle roller bearing
18 Clutch thrust piece
19 Thrust bearing
20 Thrust washer
21 Clutch release
22 Oil seal
23 Washer
24 Needle roller bearing
25 Needle roller bearing
26 Clutch operating arm
27 Bolt
28 Locating dowel

12.1 Remove the primary drive cover and the gasket

12.2a Unscrew the clutch pressure plate springs and bolts evenly

12.2b Lift out the pressure plate and then the clutch plates

12.3a Displace the tab washer on the centre nut

12.3b Use steel sprag when loosening clutch centre nut

12.4a With the centre nut removed, pull boss of shaft

12.4b Use screw to displace bearing spacer to allow ...

12.4c ... removal of the clutch drum/primary driven gear

13 Dismantling the engine unit: removing the gear selector external components and oil pump

1   Remove the circlip on the gearchange shaft left-hand side where it emerges from the engine outer casing. Grasp the gearchange shaft at the quadrant end and withdraw it from the casing, complete with the centraliser spring. Detach the change drum guide plate and the pawl operating plate, both of which are retained by two countersunk screws. Pinch together the two spring-loaded pawls which are fitted to the selector quadrant, and withdraw the quadrant, complete with pawls, from the end of the change drum. Store the pawls and springs safely to avoid loss. Removal of all these components can be carried out at this stage or later, after separation of the crankcase halves.
2   To remove the oil pump, the oil pump driven gear must first be removed. Displace the circlip which retains the gear, and slide the gear off the oil pump drive pin. Note the fitting of a washer behind the driven gear. Remove the three crosshead screws which retain the oil pump. With these removed, pull the oil pump from place. Note the two O-rings which are fitted to the rear of the pump. There are now three plates to be detached. Remove the mainshaft bearing retainer plate, which is retained by three countersunk screws, followed by the oil passage plate, fitted above the mainshaft, and similarly held by three screws. The third plate, held by four countersunk screws, is the layshaft lubrication end plate. Note that removal of the oil passage plate, whilst not being strictly necessary, can be useful to check for clogging etc. The countersunk screws which secure all the plates within the gear casing will probably be very tight. An impact driver should be used to facilitate removal.
3   Changing to the left-hand side of the gearbox; detach the retainer plate for the oil seal fitted 'alongside the final output (layshaft) shaft.

14 Dismantling the engine unit: removing the sump and oil strainer

1   Invert the engine so that the sump is facing upwards and the crankcase upper half is resting on its rear edge and on the cylinder holding studs. Loosen evenly and remove the sump retaining screws. Lift the sump away, after releasing it from the gasket, using a rawhide mallet if necessary.
2   The oil strainer is retained on the oil pick-up chamber by three screws. Remove the screws and lift the screen away.

15 Dismantling the engine unit: separating the crankcase halves

1   Slacken evenly and remove the thirteen crankcase upper half securing bolts, having returned the engine to its normal attitude. Invert the engine again so that it is resting on the cylinder holding studs and the rear edge of the upper casing. Loosen evenly and remove the nine 6 mm and twelve 8 mm securing bolts from the lower crankcase half.
2   As the gear selector drum and forks are fitted to the lower casing, it is suggested that the engine be placed so that it is resting on the lower crankcase half. This will permit the upper half to be lifted away, leaving all the main components in the lower crankcase half.
3   Separation of the crankcase halves should be carried out with care, using a rawhide mallet initially, to release the two cases from the gasket compound which was used on original assembly. **DO NOT** use levers placed between the two mating surfaces in an effort to hasten separation. Treatment of this nature will almost certainly damage the machined surfaces, causing subsequent oil leakage.
4   After separation, study the internal components carefully before continuing with the dismantling operation. This will help prevent confusion when reassembly is being carried out. Note and remove the O-ring which seats in a recess in the upper casing. Check the two location dowels for tightness. If loose, they should be removed, to avoid loss.

13.1 Remove the circlip on the left-hand end of the gearchange shaft

13.2 With the three retaining screws removed, displace the oil pump

15.4 Note and remove the O-ring in the recess in the upper casing

16 Dismantling the engine unit: removing the crankshaft and gear shafts

1   Grasp the crankshaft with both hands and lift it upwards, out of the casing, as a complete unit, together with the oil seals and the cam chain. If the crankshaft is firmly seated, use a rawhide mallet to free it from the casing. Note the main bearing outer race location pins in the upper crankcase half. If they are loose, remove them with a pair of pliers.

2   Lift out the two gearbox shafts individually, complete with pinions and seals. Note the positions of the bearing location half clips, which prevent axial movement, and prise them from position. The two shaft assemblies should be put to one side for further attention at a later stage.

17 Dismantling the engine unit: removing the gear selector internal mechanism

1   Withdraw the selector fork rods towards the primary drive side of the engine, and displace the selector forks. There are two rods, of which the front carries one fork and the rear two forks.

2   The cam stopper arm pivots on the forward rod. The arm return spring is hooked over the rear rod.

3   Remove the neutral position drum detent housing bolt and displace the detent spring and plunger. The change drum can now be pulled out of the casing towards the primary drive side of the engine.

18 Examination and renovation: general

1   Before examining the component parts of the dismantled engine/gearbox unit for wear, it is essential that they should be cleaned thoroughly. Use a paraffin/petrol mix to remove all traces of oil and sludge which may have accumulated within the engine.

2   Examine the crankcase castings for cracks, or other signs of damage. If a crack is discovered, it will require professional attention, or in an extreme case, renewal of the casting.

3   Examine carefully each part to determine the extent of wear. If in doubt, check with the tolerance figures whenever they are quoted in the text. The following Sections will indicate what type of wear can be expected and in many cases, the acceptable limits.

4   Use clean, lint-free rags for cleaning and drying the various components, otherwise there is risk of small particles obstructing the internal oilways.

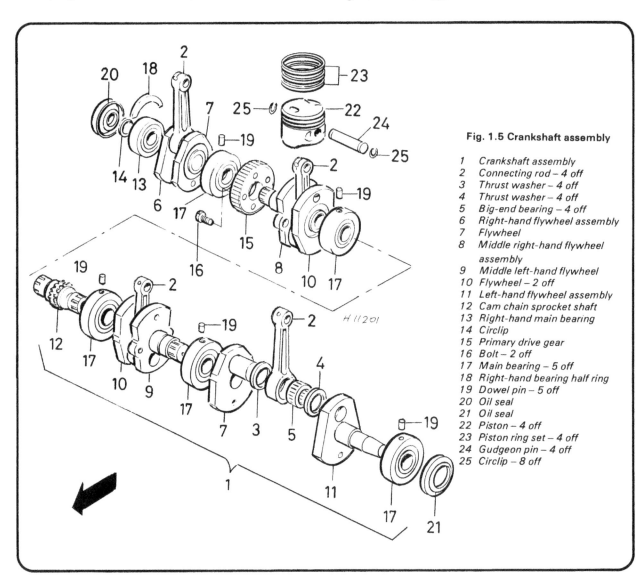

Fig. 1.5 Crankshaft assembly

1   Crankshaft assembly
2   Connecting rod – 4 off
3   Thrust washer – 4 off
4   Thrust washer – 4 off
5   Big-end bearing – 4 off
6   Right-hand flywheel assembly
7   Flywheel
8   Middle right-hand flywheel assembly
9   Middle left-hand flywheel
10 Flywheel – 2 off
11 Left-hand flywheel assembly
12 Cam chain sprocket shaft
13 Right-hand main bearing
14 Circlip
15 Primary drive gear
16 Bolt – 2 off
17 Main bearing – 5 off
18 Right-hand bearing half ring
19 Dowel pin – 5 off
20 Oil seal
21 Oil seal
22 Piston – 4 off
23 Piston ring set – 4 off
24 Gudgeon pin – 4 off
25 Circlip – 8 off

16.1 Grasp the crankshaft firmly when removing from the casing

16.2a Remove the mainshaft as a complete unit followed by ...

16.2b ... the layshaft

17.1a Remove the drum guide plate and ...

17.1b ... plunger pawl lifter plate (2 screws shown removed, arrowed)

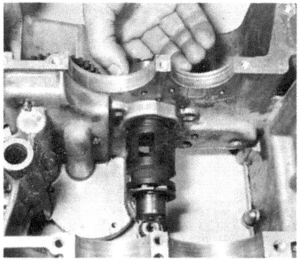

17.3 Withdraw the change drum from the casing

19 Crankshaft assembly: examination and renewal

1   The crankshaft assembly comprises four separate sets of flywheels with their respective big ends, connecting rods and main bearings, pressed together to form a single unit.

2   Due to the complex construction of the crankshaft, exchange crankshafts are not available. In the event of multiple main bearing or big-end bearing failure a new crankshaft must be acquired. If a single item only fails, it may be possible to return a crankshaft to Suzuki through a Suzuki Service Agent, for the single item to be renewed.

3   Main bearing failure will immediately be obvious when the bearings are inspected after the old oil has been washed out. If any play is evident or if the bearings do not run freely, renewal is essential. The main bearings are of the caged roller type and as the outer races are not restrained axially they may be displaced to one side, to aid visual inspection. Warning of main bearing failure is usually given by a characteristic rumble that can be readily heard when the engine is running. Some vibration will also be felt, which is transmitted via the footrests.

4   Big-end failure is characterised by a pronounced knock that will be most noticeable when the engine is working hard. There should be no play whatsoever in any of the connecting rods, when they are pushed and pulled in a vertical direction. Check also the deflection of each connecting rod in line with the crankshaft, taking the measurement at the small-end eye. Movement greater than 3 mm (0·118 inch) indicates a worn big-end bearing. Using a feeler gauge, check the axial side play at the big-ends. The total movement should be no greater than 1·00 mm (0·040 inch).

5   The oil seals at each end of the crankshaft are easy to renew when the engine is stripped; they are a push fit over each end of the crankshaft, one against and the other close to the outer main bearings. It is a wise precaution to renew these seals whenever the engine is stripped, irrespective of their condition.

20 Connecting rods: examination and renovation

1   It is unlikely that any of the connecting rods will bend during normal usage, unless an unusual occurrence such as a dropped valve has caused the engine to lock. Carelessness when removing a tight gudgeon pin can also give rise to a similar problem. It is not advisable to straighten a bent connecting rod; renewal is the only satisfactory solution.

2   The small-end eye of each connecting rod is unbushed and

it will be necessary to renew the connecting rod if the gudgeon pin becomes a slack fit. Refer to the advice about crankshaft renewal given in Section 19. Check the clearance using the unworn end of a gudgeon pin, as the centre portion of a used pin will be slightly worn. The maximum clearance should not exceed 0·09 mm (0·0035 in). Always check that the oil hole in the small-end eye is not blocked since if the oil supply is cut off, the bearing surfaces will wear very rapidly.

21 Cylinder block – examination and renovation

1   The usual indication of badly worn cylinder bores and pistons is excessive smoking from the exhausts and piston slap, a metallic rattle that occurs when there is little or no load on the engine. If the top of the bore of the cylinder block is examined carefully, it will be found that there is a ridge on the thrust side, the depth of which will vary according to the amount of wear that has taken place. This marks the limit of travel of the uppermost piston ring.

2   Measure the bore diameter just below the ridge. Take two measurements, at 90° to one another. Take two similar measurements half way down the bore and at a position just above the lower edge of the bore. If any measurement exceeds the maximum allowable, the cylinder should be rebored and fitted with an oversize piston. If the difference between the maximum and minimum measurement exceeds 0·085 mm (0·0035 in) a rebore is also required.

3   If an internal micrometer is not available, the amount of cylinder bore wear can be measured by inserting the piston without rings so that it is approximately $\frac{3}{4}$ inch from the top of the bore. If it is possible to insert a 0·060 mm (0·0024 in) feeler gauge between the piston and the cylinder wall on the thrust side of the piston, remedial action must be taken. Before going to the expense of a rebore and new pistons it is worthwhile having the exact bore dimensions checked by a specialist.

4   Oversize pistons are available in two sizes: + 0·5 mm (0·020 inch) and + 1·0 mm (0·040 in).

5   Check that the surface of the cylinder bores is free from score marks or other damage that may have resulted from an earlier engine seizure or a displaced gudgeon pin. A rebore will be necessary to remove any deep scores, irrespective of the amount of bore wear that has taken place, otherwise a compression leak will occur.

6   Make sure the external cooling fins of the cylinder block are not clogged with oil or road dirt, which will prevent the free flow of air and cause the engine to overheat.

19.3 Displace main bearing outer races and make visual inspection of bearings

19.4 Check the axial side-play at the big-ends using a feeler gauge

22 Pistons and piston rings: examination and renovation

1 Attention to the pistons and piston rings can be overlooked if a rebore is necessary, since new components will be fitted.
2 If a rebore is not considered necessary, examine each piston closely. Reject pistons that are scored or badly discoloured as the result of exhaust gases by-passing the rings.
3 Remove all carbon from the piston crowns, using a blunt scraper, which will not damage the surface of the piston. Clean away all carbon deposits from the valve cutaways and finish off with metal polish so that a clean, shining surface is achieved. Carbon will not adhere so readily to a polished surface. Using an external micrometer or vernier gauge, measure the external diameter of each piston across the thrust faces (at 90° to the gudgeon pin line), at the bottom of the skirt. Take a second measurement approximately 15 mm (0·590 in) up from the lower edge of the skirt. If either measurement is less than that given for the service limit, the piston is in need of renewal.
4 Check that the gudgeon pin bosses are not worn or the circlip grooves damaged. Check that the piston ring grooves are not enlarged. Side float should not exceed 0·18 mm (0·007 in) for the top ring and second ring, and 0·15 mm (0·006 in) for the oil control ring.
5 Piston ring wear can be measured by inserting the rings in the bore from the top, pushing them down with the base of the piston so that they are square in the bore and about 1½ inches down. If the end gap exceeds 0·65 mm (0·25 in) on any of the rings, renewal is necessary. A replacement set of rings is comparatively inexpensive and it is considered good practice to renew them as a matter of course whenever the engine is dismantled.
6 Check that there is no build up of carbon on the inside surface of the rings or in the grooves of the pistons. Any build-up should be removed by careful scraping.
7 The piston crowns will show whether the engine has been rebored on some previous occasion. All oversize pistons have the rebore size stamped on the crown. This information is essential when ordering replacement piston rings.
8 If new piston rings are fitted but a rebore has not taken place, the cylinder bores should be 'glaze busted'. This honing operation, as the name suggests, removes the highly polished glazed surface of the bore which has been caused by the countless up and down strokes of the piston and rings. If 'glaze busting' is not carried out, the time required to run-in the new rings will be greatly extended.

23 Examination and renovation: cylinder head and valves

1 Remove the cam followers and adjuster shims from the cylinder head, marking each follower so that it may be refitted in its original location. It is best to remove all carbon deposits from the combustion chambers before removing the valves for inspection and grinding-in. Use a blunt end chisel or scraper so that the surfaces are not damaged. Finish off with a metal polish to achieve a smooth, shining surface. If a mirror finish is required, a high speed felt mop and polishing soap may be used. A chuck attached to a flexible drive will facilitate the polishing operation.
2 A valve spring compression tool must be used to compress each set of valve springs in turn, thereby allowing the split collets to be removed from the valve cap and the valve springs and caps to be freed. Keep each set of parts separate and mark each valve so that it can be replaced in the correct combustion chamber. There is no danger of inadvertently replacing an inlet valve in an exhaust position, or vice-versa, as the valve heads are of different sizes. The normal method of marking valves for later identification is by centre punching them on the valve head. This method is not recommended on valves, or any other highly stressed components, as it will produce high stress

points and may lead to early failure. Tie-on labels, suitably inscribed, are ideal for the purpose. After removing each valve the valve stem oil seal should be displaced and discarded. Some difficulty may be encountered when removing a seal from the top of the guide because the shoulder on the seal is a close fit in the locating groove in the guide. Use a stout pair of long-nose pliers and wriggle the seal from place. **Warning**: when removing seals take great care not to apply excessive side load to the guides. Each guide is located longitudinally by a circlip, and the thickness of the guide wall at the circlip groove is minimal. The guides are **very** easily sheared.
3 Before giving the valve and valve seats further attention, check the clearance between each valve stem and the guide in which it operates. Clearances are as follows:

Standard	Service limit
Inlet valve/guide clearance	
0·02 – 0·05 mm	0·09 mm
(0·0008 – 0·0020 in)	(0·0035 in)
Exhaust valve/guide clearance	
0·04 – 0·07 mm	0·10 mm
(0·0015 – 0·027 in)	(0·0039 in)

Measure the valve stem at the point of greatest wear and then measure again at right-angles to the first measurement. If the valve stem diameter is below the service limit it must be renewed.

Standard	Service limit
Inlet valve stem	
6·957 – 6·975 mm	6·90 mm
(0·273 – 0·274 in)	(0·271 in)
Exhaust valve stem	
6·944 – 6·960 mm	6·80 mm
(0·273 – 0·274 in)	(0·267 in)

The valve stem/guide clearance can be measured with the use of a dial gauge and a new valve. Place the new valve into the guide and measure the amount of shake with the dial gauge tip resting against the top of the stem. If the amount of wear is greater than the wear limit, the guide must be renewed.
4 Removal of old valve guides and refitting new items should not be attempted except by a Suzuki Service Agent. The guides are a high interference fit in the cylinder head, and require special, close-fitting, pilots and drifts for successful removal and installation. In addition the bores in the cylinder head require reaming before new guides are fitted. Should valve guide renewal be necessary, return the cylinder head to a qualified agent.
5 Valve grinding is a simple task. Commence by smearing a trace of fine valve grinding compound (carborundum paste) on the valve seat and apply a suction tool to the head of the valve. Oil the valve stem and insert the valve in the guide so that the two surfaces to be ground in make contact with one another. With a semi-rotary motion, grind in the valve head to the seat, using a backward and forward action. Lift the valve occasionally so that the grinding compound is distributed evenly. Repeat the application until an unbroken ring of light grey matt finish is obtained on both valve and seat. This denotes the grinding operation is now complete. Before passing to the next valve, make sure that all traces of the valve grinding compound have been removed from both the valve and its seat and that none has entered the valve guide. If this precaution is not observed rapid wear will take place due to the highly abrasive nature of the carborundum paste.
6 If, after grinding, it is found that the width of the grey seating ring is greater than 1·5 mm (0·06 in) the valve seat must be recut using a special cutting tool. It will be seen from the accompanying illustration that angles of 75° and 15° must be cut in order to reduce the valve seat width followed by a 45° cut in order to restore the correct seat angle and the correct seat width to within the range 1·0 – 1·2 mm (0·04 – 0·05 in). Because of the expense of purchasing the three seat cutters and

because of the accuracy with which cutting must be carried out, it is strongly recommended that the cylinder head be returned to a Suzuki Service Agent for attention. It follows that when material is removed from the valve seat, the valve stem will protrude further from the upper side of the cylinder head. In extreme cases it may be found that on adjustment of the cam clearances, the prescribed clearance cannot be arrived at even with the thinnest adjustment shim available. If this is found to be the case, removal of a small amount of metal from the valve stem end is permissible. Grinding should be carried out on a suitable machine so that the stem end remains square with the shank. Where grinding to attain the correct clearance reduces the distance between the top of the stem and the upper edge of the collet groove to less than 4·00 mm (0·1574 in) a new valve seat insert must be fitted. This operation is highly skilled requiring the use of very specialised equipment.

7   Where deep pitting of the seat and valve is encountered, the seat should be recut as previously described. The valve face may be ground back on a special grinding machine to an angle of 45°, provided that after grinding, the depth of the valve periphery has not been reduced to less than 0·5 mm (0·0197 in).

8   Examine the condition of the valve collets and the groove on the valve stem in which they seat. If there is any sign of damage, new parts should be fitted. Check that the valve spring collar is not cracked. If the collets work loose or the collar splits whilst the engine is running, a valve could drop into the cylinder and cause extensive damage.

9   Check the free length of each of the valve springs. The springs have reached their serviceable limit when they have compressed to the limit readings given in the Specifications Section of this Chapter.

10 Reassemble the valve and valve springs by reversing the dismantling procedure. Ensure that all the springs are fitted with the close coils downwards towards the cylinder head. Fit new oil seals to each valve guide and oil both the valve stem and the valve guide, prior to reassembly. Take special care to ensure the valve guide oil seal is not damaged when the valve is inserted. As a final check after assembly, give the end of each valve stem a light tap with a hammer, to make sure the split collets have located correctly.

11 Check the cylinder head for straightness, especially if it has shown a tendency to leak oil at the cylinder head joint. If there is any evidence of warpage, provided it is not too great, the cylinder head must be either machined flat or a new head fitted. Most cases of cylinder head warpage can be traced to unequal tensioning of the cylinder head nuts and bolts by tightening them in incorrect sequence.

22.4 Using a feeler gauge, check the wear on the piston ring grooves and ...

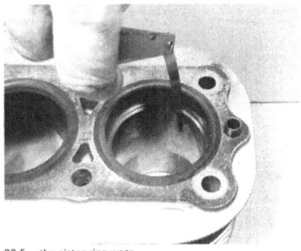

22.5 ... the piston ring wear

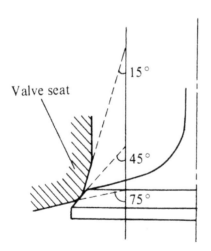

Fig. 1.6 Valve seat re-cutting angles

23.1 Valve train components, general view

23.10a Fit new oil seals to each valve guide and ...

23.10b ... ensure they are seated properly and not damaged

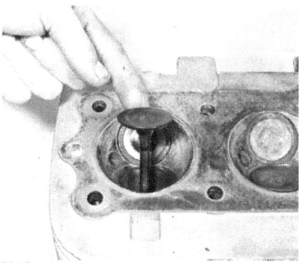

23.10c Lubricate the valve stem thoroughly before insertion

23.10d Compress the springs and refit the spring collets

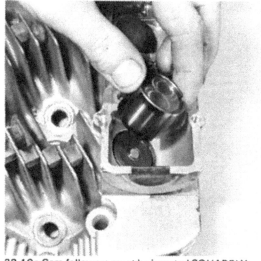

23.10e Cam followers must be inserted SQUARELY

23.10f Cam clearance adjustment shims are marked for identification (arrowed)

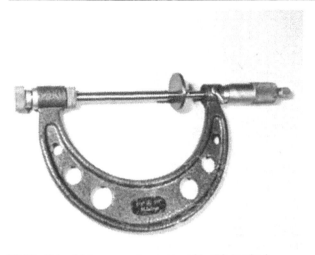

23.10g Shim thickness can be measured by this method

23.10h Replace shim of correct size to obtain correct valve/cam clearance

24 Examination and renovation: camshafts, cam followers and camshaft drive sprockets

1   Inspect the cams for signs of wear such as scored lobes, scuffing, or indentation. The cams should have a smooth surface. The complete camshaft must be replaced if any lobes are worn or indented, through lubrication failure etc. In due course even normal wear of each cam lobe may progress to the stage where full valve lift is no longer possible. Measure each cam from the lobe to the base circle, comparing the overall height with these figures.

Minimum cam height:
   Inlet   *36·00 mm (1·417 in)*
   Exhaust   *35·50 mm (1·397 in)*

If either of the camshafts has any cam below the minimum figure, that camshaft should be renewed in order to restore performance.
2   Refit both camshafts in the cylinder head and fit the bearing caps and bolts. Tighten the bolts to a torque wrench setting of 0·8 – 1·2 kgf m (6 – 8 lbf ft). Check the clearance between the camshaft journals and the bearing surfaces. This is most easily accomplished by fitting a dial gauge to the cylinder head and moving the camshaft radially in a vertical or horizontal plane. If the clearances exceed those figures given in the specifications, remove the camshafts, refit the bearing caps and check the diameter of each bearing, to determine whether the camshaft or cam bearing is at fault.
3   If it is found that the camshaft bearings are worn or badly scored, the cylinder head and bearing caps must be renewed. There is no provision for renewing the bearings as the camshafts run directly in the cylinder head material.
4   Examine the camshaft chain sprockets for hooked, worn, or broken teeth. If any damage is found, the camshaft sprocket in question should be renewed. Each sprocket is retained on the camshaft flange by two socket screws. When refitting either sprocket note that each is marked IN or EX as are the camshafts. It is important that the sprockets are fitted on the correct camshaft and in the position shown in the accompanying illustration. Incorrect assembly will prevent accurate valve timing. Apply a small quantity of locking fluid to the securing screws during reassembly.

5   The camshaft drive sprocket is an integral part of the crankshaft and therefore if damage is evident, the crankshaft must be renewed. Fortunately, this drastic course of action is rarely necessary since the parts concerned are fully enclosed and well lubricated, working under ideal conditions.
6   Inspect the external surfaces of the cam followers for signs of scoring or fracture. If damage is evident, the component must be renewed. If scoring has occurred it follows that similar damage may be found in the appropriate guide tunnel in the cylinder head. Damage to the tunnels cannot be rectified under normal circumstances and therefore a new cylinder head must be obtained. Insert each cam follower into the guide tunnel from which it was removed. Only the lightest pressure should be used to insert each follower. If a follower is inserted even at a slight angle, binding against the tunnel will result. Any effort made to tap the follower in will almost certainly jam the follower solidly. Removal is then very difficult! Check the clearance between each cam follower and guide tunnel. Unfortunately no precise figures are available but the follower should be a good sliding fit, with no perceptible play from side to side. Excess play will allow the cam follower to tilt, causing noisy operation and accelerated wear of the cylinder head.

24.4a Inlet camshaft is marked accordingly as is ...

24.4b ... the exhaust camshaft to aid reassembly

24.4c Camshaft sprockets are also marked to avoid confusion when rebuilding (EX mark arrowed)

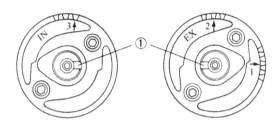

Fig. 1.7 Cam sprocket position on camshaft

25 Examination and renovation: cam chain and chain tensioner mechanism

1   Inspect the cam chain for obvious signs of damage, such as broken or missing rollers or fractured links. Some indication of the amount of chain wear may be gained by checking the extent of adjustment remaining on the automatic tensioner assembly. If the plunger has moved towards the end of the stroke, it may be assumed that the chain is near the end of its useful life. Wear of the chain can be measured by washing it in petrol, then pulling on the chain so that it stretches as far as possible. If the length of 20 links exceeds 157.8 mm, the chain must be renewed. Although the cam chain works in almost ideal conditions, being fully lubricated and enclosed, wear will develop after an extended mileage. If there is any doubt as to the chain's condition, it should be renewed, as breakage will cause extensive engine damage.

2   Loosen the locking screw on the chain tensioner body to free the plunger pushrod. Rotate the adjuster knob anticlockwise so that the plunger may be pushed in'fully, and check that the plunger moves in and out freely, without any tendency to bind. If plunger movement is not perfectly smooth, the complete unit should be renewed.

3   Inspect the surfaces of the two chain guide blades. If the rubber has been badly scored by the chain or is coming away from the steel backing, the blade in question should be renewed. The rubber bridge guide between the two camshaft sprockets should also be inspected for wear, and again, replaced if necessary.

26 Examination and renovation: tachometer drive assembly

1   The worm drive to the tachometer is an integral part of the exhaust camshaft which meshes with a pinion attached to the cylinder head cover. If the worm is damaged or badly worn, it will be necessary to renew the camshaft complete.

2   The driveshaft and pinion are a single part retained in the cylinder head in a bush housing which is secured by a claw and screw. Renewal is therefore straightforward. It is unlikely that wear will develop on either the drive or driven pinion as both are well lubricated and lightly loaded.

27 Examination and renovation: gearbox components

1   It should not be necessary to dismantle either of the gear clusters unless damage has occurred to any of the pinions or if the caged needle roller bearings require attention.

2   The accompanying illustration shows how both clusters of the gearbox are assembled on their respective shafts. It is imperative that the gear clusters, including the thrust washers, are assembled in EXACTLY the correct sequence, otherwise constant gear selection problems will occur.

In order to eliminate the risk of misplacement, make rough sketches as the clusters are dismantled. Also strip and rebuild as soon as possible to reduce any confusion which might occur at a later date.

3   When dismantling the gear shafts, the journal ball bearings may be pulled from position, using a standard two or three-legged sprocket puller. The layshaft right-hand bearing should be removed by placing the puller on the 1st gear pinion and drawing the two components off simultaneously. Care should be taken here to ensure that the bearing retaining C-clips, if not already removed when the crankcases were separated, are displaced and stored safely for use during reassembly. A dogged washer is fitted between the 3rd and 4th gear pinions and 2nd and 5th gear pinions on the layshaft. Between the 3rd gear and 5th gear pinions, on the layshaft, is another dogged washer, and a second, special, lock washer. To free this locking washer, and release the next washer in the sequence, it must be turned slightly so that the internal serrations clear the shaft splines. The washer may then be pulled from place. Similarly, on reassembly, the special washer should be turned so that it engages correctly. Ensure that the internally spline sleeve which fits against the lock washer is fitted so that the oil hole aligns with the hole in the shaft.

4   The 2nd gear pinion on the mainshaft is an interference fit
and will require pulling from position. On refitting this pinion, it
is essential that it is so placed that the distance between its
outer face and that of the 1st gear pinion is 109·4 mm – 109·5
mm (4·307 in – 4·311 in). This measurement is critical for
correct rotational clearance and perfect alignment. Before refitting the pinion, treat the inner bore with a high shear strength
locking compound. Suzuki recommend the use of Thread Lock
Super 103K, although in the absence of this, one of the more
easily obtainable, good quality locking compounds, would be an
acceptable substitute. After refitting the 2nd gear pinion, check
that the 5th gear pinion is free to rotate, and has not become
locked by excess locking fluid.

5   Give the gearbox components a close visual inspection for
signs of wear or damage such as broken or chipped teeth, worn
dogs, damaged or worn splines and bent selectors. Replace any
parts found unserviceable because they cannot be reclaimed in
a satisfactory manner.

6   The gearbox bearings must be free from play and show no
signs of roughness when they are rotated. After thorough
washing in petrol the bearings should be examined for
roughness and play. Also check for pitting on the roller tracks.

7   It is advisable to renew the gearbox oil seals irrespective of
their condition. Should a re-used oil seal fail at a later date, a
considerable amount of work is involved to gain access to
renew it.

8   Check the gear selector rods for straightness by rolling
them on a sheet of plate glass. A bent rod will cause difficulty in
selecting gears and will make the gear change particularly
heavy.

9   The selector forks should be examined closely, to ensure
that they are not bent or badly worn. The pegs which engage
with the cam channels are integral with the forks therefore if
they are worn the forks must be renewed. Under normal conditions, the gear selector mechanism is unlikely to wear quickly,
unless the gearbox oil level has been allowed to become low.

10   The tracks in the selector drum, with which the selector
forks engage, should not show any undue signs of wear unless
neglect has led to under lubrication of the gearbox. Check the
condition of the gearchange arm and drum stopper arm springs.
Weakness in the springs will lead to imprecise gear selection.
Although unlikely to show signs of wear before a considerable
mileage has been recorded, the condition of the gear stopper
arm roller should be checked.

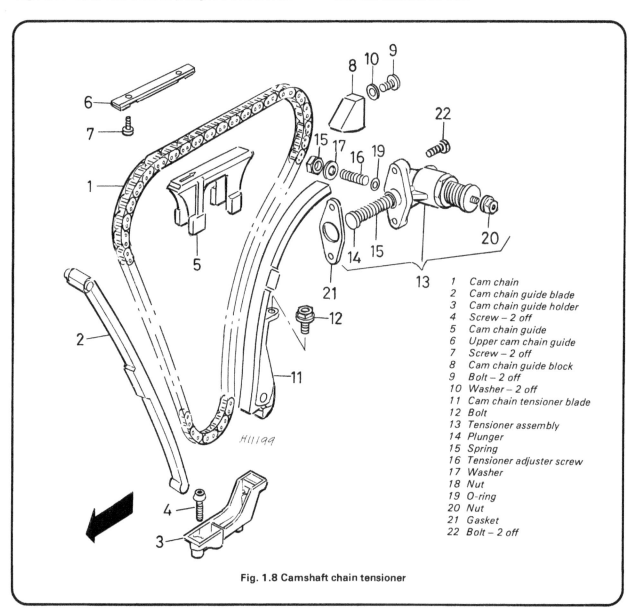

1   Cam chain
2   Cam chain guide blade
3   Cam chain guide holder
4   Screw – 2 off
5   Cam chain guide
6   Upper cam chain guide
7   Screw – 2 off
8   Cam chain guide block
9   Bolt – 2 off
10 Washer – 2 off
11 Cam chain tensioner blade
12 Bolt
13 Tensioner assembly
14 Plunger
15 Spring
16 Tensioner adjuster screw
17 Washer
18 Nut
19 O-ring
20 Nut
21 Gasket
22 Bolt – 2 off

Fig. 1.8 Camshaft chain tensioner

Fig. 1.10 Installation distance when fitting 2nd gear pinion to mainshaft

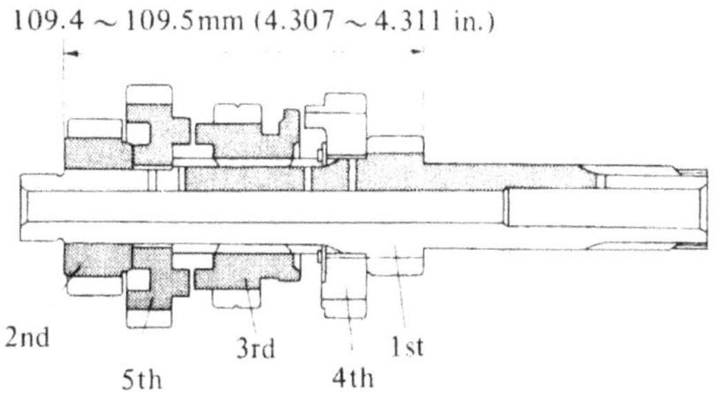

109.4 ~ 109.5mm (4.307 ~ 4.311 in.)

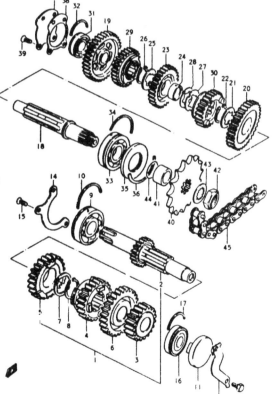

2nd 5th 3rd 1st 4th

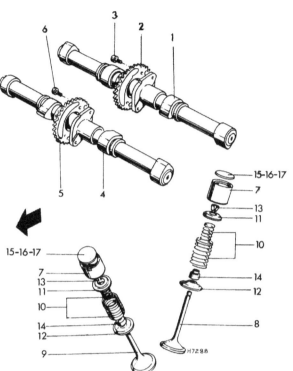

Fig. 1.9 Camshafts and valves

1 Inlet camshaft	10 Valve spring set – 8 off
2 Inlet cam sprocket	11 Spring collar – 8 off
3 Bolt – 2 off	12 Spring seat – 8 off
4 Exhaust camshaft	13 Collet – 16 off
5 Exhaust cam sprocket	14 Oil seal – 8 off
6 Bolt – 2 off	15 Adjuster pad – AR
7 Camfollower – 8 off	16 Adjuster pad – AR
8 Inlet valve – 4 off	17 Adjuster pad – AR
9 Exhaust valve – 4 off	

Fig. 1.11 Gearbox components

1 Mainshaft assembly	24 Bush
2 Mainshaft	25 Splined tab washer
3 Mainshaft – 2nd gear pinion	26 Circlip
4 Mainshaft – 3rd gear pinion	27 Splined lock washer
5 Mainshaft – 4th gear pinion	28 Special splined washer
6 Mainshaft – 5th gear pinion	29 Layshaft – 4th gear pinion
7 Splined thrust washer	30 Layshaft – 5th gear pinion
8 Circlip	31 Layshaft right-hand bearing
9 Journal ball bearing	32 Half ring
10 Half ring	33 Layshaft left-hand bearing
11 End cap	34 Half ring
12 End cap retaining plate	35 Oil seal
13 Bolt – 2 off	36 Half ring
14 Bearing retainer	37 Blanking plate
15 Screw – 3 off	38 Gasket
16 Mainshaft left-hand bearing	39 Screw – 4 off
17 Half ring	40 Final drive sprocket
18 Layshaft	41 Spacer
19 Layshaft – 1st gear pinion	42 Nut
20 Layshaft – 2nd gear pinion	43 Splined tab washer
21 Splined thrust washer	44 O-ring
22 Circlip	45 Final drive chain
23 Layshaft – 3rd gear pinion	

28 Examination and renovation: clutch assembly

1   After an extended period of use the clutch linings will wear and promote clutch slip. The clutch plates should be measured with a vernier gauge or pair of calipers to ascertain the extent of wear. The measurement of the thickness for the inserted (friction) plates and the maximum wear limits are as follows:

Inserted plate thickness 2·9 – 2·7 mm (0·11 – 0·10 in)
Wear limit   2·5 mm (0·09 in)

If the plates thickness is less than the specified minimum they must be renewed.
2   The plain clutch plates should not show any evidence of overheating (blueing). If they do, check them for overall flatness by placing each plate on a flat surface and measuring the bow with a feeler gauge. If any of the plates are warped by more than 0·1 mm (0·004 in) they should be renewed.
3   Check the free length of each clutch spring. If the springs have shortened (set) to a length less than the specified minimum, set out below, they must be renewed. Renew the springs as a set, rather than individually.

Clutch spring free length 40·5 – 39·0 mm (1·59 – 1·53 in)
Wear limit   38·5 mm (1·51 in)

4   Check the condition of the clutch centre spacer and the external caged needle roller bearing. If wear is evident in these components, they should be renewed. The bearing and spacer upon which the oil pump drive gear (fitted behind the clutch) is mounted should be checked similarly.
5   Check the condition of the slots in the outer surface of the clutch centre and the inner surfaces of the outer drum. In an extreme case, clutch chatter may have caused the tongues of the inserted plates to make indentations in the slots of the outer drum, or the tongues of the plain plates to indent the slots of the clutch centre. These indentations will trap the clutch plates as they are freed and impair clutch action. If the damage is only slight the indentations can be removed by careful work with a file and the burrs removed from the tongues of the clutch plates in similar fashion. More extensive damage will necessitate renewal of the parts concerned.
6   Check the clutch release thrust bearing in the pressure plate. If play is evident or the bearing rotates roughly, it should be renewed.
7   Visually check the condition of the shock absorber springs fitted to the rear of the clutch outer drum/primary drive gear assembly. With the high power output of this machine, these springs are frequently subjected to harsh treatment during acceleration. Check for scoring marks on the springs where they have been moving, and note if there is movement promoted by shaking the unit; badly worn springs will be obvious by a pronounced rattle, when this is done. If wear is excessive then the whole unit must be renewed.
8   The clutch release mechanism in the clutch cover does not normally require attention, provided it is greased from time to time. If the unit fails, it must be renewed as a complete assembly.

29 Crankcase covers: examination and renovation

1   The right-hand and left-hand crankcase cover and the inspection covers are unlikely to become damaged unless the machine is dropped or involved in an accident. Cracks in a casing can be repaired easily by special aluminium welding, providing the damage is not too extensive and care is taken to prevent distortion.
2   The covers are lightly polished and lacquered before leaving the factory. Badly scratched covers can be refurbished using a single cut file treated with chalk to prevent clogging, and finished off with fine emery paper and metal polish or aluminium cleaner. If required, the cases can be relacquered, using an aerosol paint spray.

30 Engine reassembly: general

1   Before reassembly of the engine/gear unit is commenced, the various component parts should be cleaned thoroughly and placed on a sheet of clean paper, close to the working area.
2   Make sure all traces of old gaskets have been removed and that the mating surfaces are clean and undamaged. One of the best ways to remove old gasket cement is to apply a rag soaked in methylated spirit. This acts as a solvent and will ensure that the cement is removed without resort to scraping and the consequent risk of damage. If a gasket becomes bonded to the surface through the effects of heat and age, a new sharp scalpel blade should be used to effect removal. Old gasket compound can also be removed using a soft brass wire brush of the type used for cleaning suede shoes. A considerable amount of scrubbing can take place without fear of damaging the mating surfaces.
3   Gather together all the necessary tools and have available an oil can filled with clean engine oil. Make sure that all new gaskets and oil seals are to hand, also all replacement parts required. Nothing is more frustrating than having to stop in the middle of a reassembly sequence because a vital gasket or replacement part has been overlooked.
4   Make sure that the reassembly area is clean and that there is adequate working space. Refer to the torque and clearance settings wherever they are given. Many of the smaller bolts are easily sheared if overtightened. Always use the correct size screwdriver or bit for the crosshead screws never an ordinary screwdriver or punch. If the existing screws show evidence of maltreatment in the past, it is advisable to renew them as a complete set.

31 Engine reassembly: replacing the gear change drum and internal selector components

1   If the change drum stopper plate was removed from the drum for inspection or renewal, it must be refitted at this stage. The plate must be positioned so that it locates correctly with the drive pin which is a push fit in the drum end boss. Secure the plate by means of the circlip. Assemble the gear selector quadrant together with the two selector pawls, the plugers and the springs. The pawls must be fitted so that the narrower edges adjacent to the plunger recesses face towards the rear of the selector quadrant. Depress the pawls against the spring pressure and insert the completed selector quadrant into the end of the change drum.
2   Lubricate the change drum needle roller bearing with engine oil, and slide the drum into position in the gearbox. Refit the neutral stopper plunger, detent spring and the bolt.
3   Slide the selector fork rods into position through the right-hand gearbox wall and refit the selector forks. The two rearmost selector forks, which share the same rod, are identical, and are fitted with their guide pins uppermost. Check that the forks are positioned the correct way round. Slide the change drum stopper arm onto the front selector rod. With the components correctly in place push both rods fully home and reconnect the stopper arm spring with the rear selector fork rod.
4   Coat the four countersunk screws which retain the change drum retainer plate, and the pawl plate, with locking fluid, and replace the two plates. Rotate the change drum until it is in the neutral position and the neutral stopper locates.

31.2a Lubricate the change drum needle roller bearing before fitting the change drum

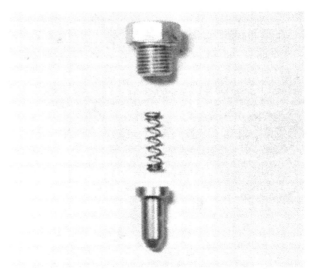

31.2b Detent spring, bolt, and neutral stopper plunger (general view)

31.2c Fit the detent bolt and spring

31.3a Slide in the rear selector rod and two selectors

31.3b Install forward selector rod to secure the fork and arm

31.3c Tension the stopper and spring so it anchors on the rear selector rod

32 Engine reassembly: replacing the gear shaft assemblies

1   Position the upper crankcase half so that it rests on the cylinder holding the studs and the rear of the casing. Before refitting, the gearshafts must be assembled as completed sub-assemblies, including the gear pinions bearings and oil seals. The oil seal lips should be lubricated before being installed to prevent damage.

2   Install the bearing securing half C-clips in the casing grooves and lower the gear shafts into place. It will probably prove easier to fit the shafts individually, but they can be fitted as a meshed pair if it is so desired. Three of the four bearings are fitted with a single location pin each.

3   Replace the end plate and oil seal at the blind end of the mainshaft. The end plate should be positioned with the raised dome against the bearing outer races so that there is a gap between the plate and bearing. The oil seal on the layshaft output end is supported by a crescent shaped plate which locates with a groove in the upper casing.

33 Engine reassembly: replacing the crankshaft

1   Lubricate the main bearings thoroughly and also the crankshaft ends onto which are to be fitted the oil seals. The crankshaft right-hand seal should be fitted so that the cupped side is facing outwards, with the projections facing inwards in contact with the bearing. The left-hand seal is fitted with the spring garter side facing inwards.

2   Insert the five main bearing locating pegs into the holes provided in the casing. The right-hand main bearing is located by a half clip in a manner similar to that of the gearbox ball bearings. In addition a small dowel is fitted, which on installation of the crankshaft must be positioned in the recess adjacent to the bearing housing.

3   Fit the cam drive chain over the crankshaft so that it meshes with the drive sprocket. Grasp the crankshaft at both ends and lower the completed assembly into position. Ensure that the main bearing outer races engage correctly with the locating dowels. A tiny punch mark is provided on each main bearing outer race, diametrically opposed to the centre of the dowel hole. Lining up all five marks in a perfectly straight line will aid correct location of the bearings. To aid the seating of each bearing firmly in its bearing housing, the gentle use of a rawhide mallet is permissible. This should, of course, only be attempted after ensuring correct location of the locating pegs and the half clip.

34 Engine reassembly: joining the crankcase halves

1   Carefully clean the crankcase halves mating surfaces. Replace the two hollow locating dowels, tapping them into position carefully so as not to distort them. If the dowels have become slightly burred they should be cleaned up with a small file.

2   Smear the upper crankcase half mating face with a thin layer of jointing compound. Suzuki recommend that Suzuki Bond No. 4 be used to make the joint. A good quality non-hardening compound will make a suitable substitute. Fit the small O-ring into the recess to the front of the mainshaft assembly.

3   Let the gasket compound set for at least 10 minutes and then lower the upper casing down into place. No difficulty should be encountered in fitting the upper case but special care should be taken that the three selector forks locate with their respective guide ways in the sliding pinions. To aid this operation, ensure that the change drum is in the neutral position and arrange the gears so that they too are in neutral. When the lower casing is lowered into position, the selector forks fall naturally into a vertical position and so engage easily with the pinions.

4   Before refitting the crankcase bolts a check should be made on the operation of the reassembled gearbox components. Temporarily refit the gearchange lever and ensure that all five gears are engaging and disengaging in the correct manner. If the operation is not correct initially, the selector forks may not have engaged exactly with their respective guide ways. Care and patience may be required in order to manoeuvre the components into their respective positions. Remove the gearchange lever when the check has proved the operation of the gearbox to be correct.

5   Fit the crankcase retaining bolts to the lower crankcase. Tighten the 8 mm bolts evenly a little at a time. Note that there are two bolts which carry clips for the alternator cable; these should be fitted to the right-hand side of the casing. The numerical sequence shown in the accompanying illustration should be followed during the tightening procedure. Tighten the nine 6 mm bolts, again using the numerical sequence. Invert the crankcase and fit the upper bolts. All crankcase bolts should be tightened to the torque settings specified below:

Crankcase bolts

8 mm	2·0 kgf m (14·5 lbf ft)
6 mm	1·0 kgf m (7·2 lbf ft)

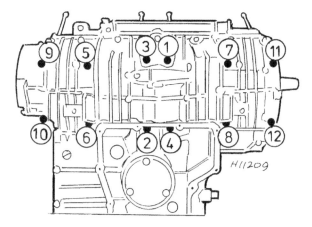

Fig. 1.12 Crankcase bolt tightening sequence

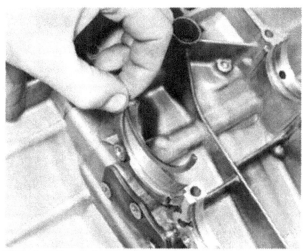

32.2a Install mainshaft bearing half C-clips and ...

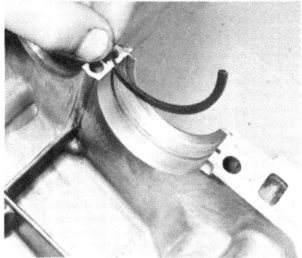

32.2b ... the similar clips to retain the layshaft bearings

32.2c Install layshaft followed by ...

32.2d ... the complete mainshaft

32.2e Arrange bearing so that pin locates in recess

32.3 Replace oil seal on mainshaft as shown

33.1 Lubricate the main bearings thoroughly

33.2a Install the five main bearing pegs and ...

33.2b ... the right-hand main bearing half-clip

33.3 Mesh the camshaft drive chain over crankshaft drive sprocket before fitting crankshaft assembly

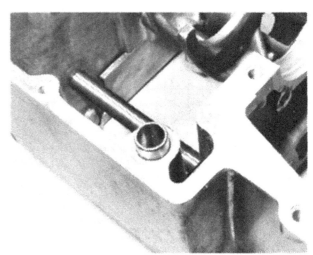

34.1 Replace hollow locating dowels

34.3 Lower the crankcase lower half into position

35 Engine reassembly: replacing the oil strainer screen, sump and oil filter

1   Place the oil strainer screen in position on the underside of the pick-up chamber and fit the retaining screws. A little locking fluid may be used on the screws, to ensure that they cannot become loose.

2   Fit a new sump gasket followed by the sump and retaining bolts. No gasket cement is required at this joint. Tighten the screws evenly, to avoid distortion. Fit and tighten the oil drain plug, ensuring that the plug is not overtightened. With frequent recommended oil changes the thread can become worn. Too much force will strip the thread.

4   Place a new gasket over the oil filter chamber studs and insert the oil filter element with the O-ring end facing inwards. Position the coil spring against the filter and then replace the cover. Fit and tighten the three domed retaining nuts.

35.1 Apply locking fluid to oil strainer screen screws

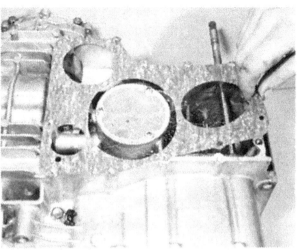

35.2a Fit new gasket before ...

35.2b ... refitting the sump base plate

36 Engine reassembly: replacing the gear selector external components

1 Refit the mainshaft bearing retainer plate in the primary chain case, and in addition, the layshaft blind end plate and the lubrication passage end plate. Apply locking fluid to the thread of all these screws, to ensure security.

2 Grease the gear change shaft oil seal on the left-hand side of the gearbox. Insert the change shaft, complete with the main change arm, into position, taking care that the splines on the shaft end do not damage the oil seal. Mesh the teeth on the change arm with the teeth on the change drum, as shown in the accompanying photograph. The change arm centraliser spring must be fitted on the arm as shown. When in position the spring arms must lie either side of the anchor peg in the casing.

3 Fit the washer and spring circlip to the left-hand end of the gearchange shaft, to secure it firmly in place.

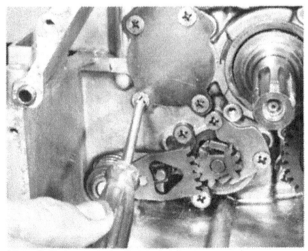

36.1a Refit the layshaft blind end plate screws and ...

36.1b ... those on the mainsaft bearing retainer plate with locking fluid applied

36.2 Install change arm to mesh with quadrant and ensure change arm centraliser spring is correctly fitted (arrowed)

37 Engine reassembly: replacing the oil pump and oil pump drive gear

1   The oil pump must be re-installed in the primary drive casing as a complete unit, either before or after the driven gear is fitted. The driven gear is retained by a circlip on the shaft end, and is located by a drive pin which passes through the shaft, engaging with a recess in the rear face of the pinion.
2   Place a new O-ring in each of the casing recesses against the wall of which the pump is secured. Omission of the O-rings will lead to lubrication failure. Position the oil pump and fit and tighten the three mounting screws.
3   The oil pump drive gear assembly can now be replaced. Fit the heavy backing washer onto the clutch shaft, followed by the drive gear bearing spacer and needle roller bearing. Lubricate

the bearing thoroughly and then refit the drive pinion. The pinion must be fitted with the two projecting dogs facing outwards, as they locate with the rear of the clutch drum, so providing the driving operation.

38 Engine reassembly: replacing the clutch

1   Place the clutch outer drum over the clutch shaft (mainshaft) and into the casing. Centralise the drum and after lubrication, fit the spacer and needle bearing. The spacer should be fitted with the radially grooved face inwards. When fitting the outer drum, it is essential that the two projecting dogs on the oil pump drive gear engage with two recesses in the drum rear face. Insert a narrow shanked screwdriver through one of the two holes in the clutch spacer so that the drive pinion is prevented from rotating. Turn the clutch outer drum until it can be felt that the pinion and drum have engaged correctly.
2   Install the heavy thrust washer on the clutch shaft, with the milled channel facing inwards, and then refit the clutch centre boss onto the clutch shaft splines. Fit the tab washer and then fit and tighten the centre nut. Use the same procedure for tightening the nut as used for loosening.See Section 12, paragraph 3, of this Chapter. After tightening the nut, do not omit to bend up the tab washer to secure the nut in place.
3   Replace the clutch plates one at a time, commencing with a friction (inserted) plate followed by a plain plate and so on, alternately. Lubricate the clutch thrust piece and fit it, together with the thrust bearing and shim. Refit the clutch pressure plate and the clutch springs, washers and bolts. Tighten the bolts fully.

Fig. 1.13 Gearchange mechanism

1   Layshaft selector fork – 2 off
2   Mainshaft selector fork
3   Selector fork rod – 2 off
4   Gear change drum
5   Selector quadrant
6   Change pawl
7   Change pawl
8   Plunger – 2 off
9   Spring – 2 off
10   Pawl lifter plate
11   Screw – 2 off
12   Guide plate
13   Screw – 2 off
14   Shim
15   Bearing
16   Stopper arm
17   Spring
18   Detent plunger
19   Housing bolt
20   Spring
21   Sealing washer
22   Stopper plate
23   Pin
24   Circlip
25   Change shaft
26   Centraliser spring
27   Stopper bolt
28   Washer
29   Oil seal
30   Washer
31   Circlip
32   Gearchange pedal
33   Rubber
34   Pinch bolt
35   Gear position switch
36   O-ring
37   Screw – 2 off
38   Washer – 2 off
39   Washer – 2 off

37.1a Fit the two O-rings into the casing before ...

37.1b ... installing the oil pump

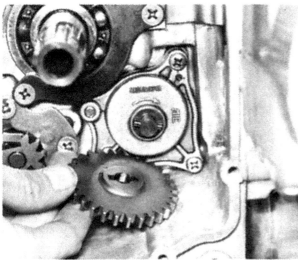

37.1c Oil pump gear is located by a drive pin and ...

37.1d ... secured by a small circlip

37.3 Fit oil pump drive pinion and bearing, noting two projections which face outwards (arrowed)

38.1a Install the clutch drum, ensuring the pump gear projections locate in the recesses in the clutch drum rear

38.1b Lubricate and refit clutch bearing and spacer

38.2 Fit clutch centre boss and ...

38.3a ... lockwasher and centre nut, then replace plates alternately, one at a time

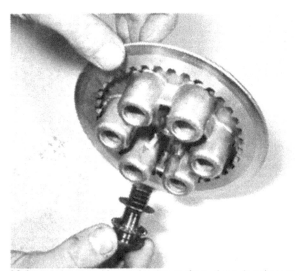

38.3b Lubricate and insert the thrust piece, thrust bearing and washer

38.3c Replace the clutch pressure plate and ...

38.3d ... fit the clutch springs, washers and bolts

38.3e Tighten the bolts fully

39 Engine reassembly: refitting the primary drive cover

1 The primary drive cover can now be replaced. Lubricate the primary drive gears with engine oil and then fit a new gasket to the mating surface. Push the cover into position on the two hollow locating dowels and then fit the screws. The screws should be tightened evenly, in a diagonal sequence, to help prevent distortion. Care should be taken to ensure that the teeth on the clutch release thrust piece engage with those on the release shaft pinion.

40 Engine reassembly: replacing the starter motor clutch and intermediate gear and alternator

1 Place the starter clutch pinion bearing thrust washer onto the left-hand end of the crankshaft so that the face with chamfered inner radius is towards the main bearing, and the four notches face outwards. Lubricate the double needle roller bearings and fit them, together with the clutch pinion.

2 Before fitting the combined alternator rotor/starter clutch unit, clean thoroughly the external taper on the crankshaft end and the internal taper in the alternator rotor. Position the rotor/clutch assembly on the shaft, turning it anti-clockwise so that the three clutch rollers slide easily onto the pinion boss. Insert and tighten the rotor centre bolt. The correct torque wrench setting is 9·0 – 10·0 kgf m (65·0 – 72·3 lbf ft).

39.1 Replace primary drive cover, ensure teeth on release thrust piece engage on release shaft pinion (arrowed)

3 Place the intermediate double gear on its stub shaft and fit one thrust washer either side of the gear. Position the assembly in the casing with the larger gear to the rear of the starter clutch pinion, and push the stub shaft fully home into the casing recess.

4 Fit a new gasket to the crankcase surface and place the alternator cover close to the engine so that the alternator leads may be threaded through the wall of the starter clutch chamber into the starter motor compartment. Pull the wires through whilst fitting the outer cover. Fit and tighten evenly the casing holding screws.

41 Engine reassembly: replacing the neutral indicator switch

1 Before refitting the switch, replace the oil seal retainer plate which is fitted to the outside of the left-hand gearbox wall. The tongue projecting from the plate must be fitted uppermost, as it serves as a guide for the neutral indicator (or gear position) switch lead.

2 Position the O-ring in the switch recess and insert the contact spring in the hole in the change drum end. Replace the neutral indicator body and tighten down the two retaining bolts. Route the wire through the guide channel provided and restrain it additionally, by means of the oil seal plate tongue.

42 Engine reassembly: replacing the ATU and contact breaker assembly

1 Position the ATU against the end of the crankshaft so that the drive pin projecting from the shaft end engages with the recess in the rear of the unit. Fit the timing index pointer plate into the casing so that scribed mark is at the 12 o'clock position, and then install the complete contact breaker assembly stator plate. The stator plate should be fitted so that the wiring lead grommet can locate with the rebated hole in the casing wall.

2 If, on dismantling, a punch or scratch mark was made on the stator plate aligning with a similar mark on the casing, these marks should be realigned and the three retaining screws refitted. It is probable that the ignition timing will be correct but a check must be made as a precautionary measure. Where no alignment marks were made, ignition timing should be set as a matter of course, as described in Chapter 3, Section 7.

3 Before checking or setting the ignition timing, replace the ATU centre bolt and the hexagonal engine turning piece. To tighten the bolt, pass a close fitting bar through one small-end eye, bearing down on two wooden blocks placed across the crankcase mouth. Do not fit the contact breaker cover before fitting and timing the valves.

40.1a Fit starter clutch pinion bearing thrust washer with notches outermost and ...

40.1b ... lubricate and fit double needle roller bearings

40.2 Fit and tighten the centre bolt

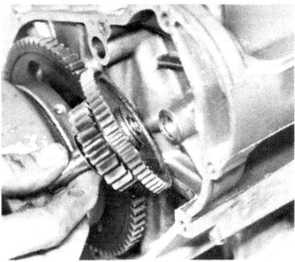
40.3 Fit thrust washer each side of intermediate gear

41.1 Replace the oil seal plate with the tongue upwards

41.2a Fit the contact brush and spring

41.2b Refit switch body, not forgetting the O-ring

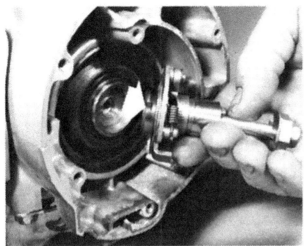

42.1a Install ATU so that notch (arrowed) ...

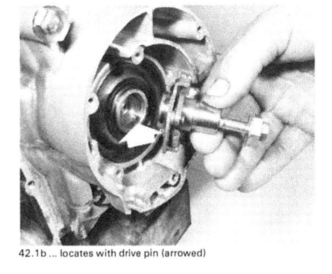

42.1b ... locates with drive pin (arrowed)

42.1c Fit index plate in position shown

42.1d Re-align scratch marks made on dismantling (arrowed) and feed wiring lead through casing grommet

43 Engine reassembly: replacing the pistons and the cylinder block

1   Fit the piston rings to each piston, commencing with the three-piece oil control ring. The first of the three parts to be fitted should be the corrugated spacer band. Fit the oil control ring side rails one at a time. When fitting the two compression rings, ensure that both are fitted with the R mark facing upwards. Do not mix rings of different makes in the same engine. The two compression rings are of differing type and cross-section. The upper ring has a chrome plated face which is slightly curved. The 2nd ring is not chrome plated and has a tapered face. Before replacing the pistons, pad the mouths of the crankcase with rag in order to prevent any displaced component from accidentally dropping into the crankcase.

2   Fit the pistons in their original order with the arrow on the piston crown pointing toward the front of the engine.

3   If the gudgeon pins are a tight fit, first warm the pistons to expand the metal. Oil the gudgeon pins and small end bearing surfaces, also the piston bosses, before fitting the pistons.

4   Always use new circlips **never** the originals. Always check that the circlips are located properly in their grooves in the piston boss. A displaced circlip will cause severe damage to the cylinder bore, and possibly an engine seizure.

5   Replace the chain tensioner rear guide blade, ensuring that the lower end locates correctly with the guide seat retained in the crankcase. Fit and tighten the single securing bolt. Install the two sealing rings in the recesses surrounding the outer rear cylinder studs and then fit a new cylinder base gasket over the studs. Check that the two hollow locating dowels are fitted to the outer front holding down studs. A new O-ring should be placed on each of the four cylinder bore spigots. Push the-O rings fully home, so that they seat correctly in the grooves provided.

6   Arrange the piston ring gaps as shown in the accompanying diagram, in order to maintain the best sealing characteristics. Using clean engine oil, lubricate thoroughly the cylinder bores. Lift the cylinder block up onto the studs and support it there whilst the camshaft chain is threaded through the tunnel between the bores.

7   On a multi-cylinder engine of this type where the pistons must be introduced into the bores simultaneously, the use of piston ring compressors or clamps is an advantage. Each cylinder bore, however, is provided with a generous lead-in which will ease hand feeding of the pistons in the absence of compressors. Gently lead the pistons into the bores, working across from one side. It is obviously an advantage here to have at least one other pair of hands to assist the operator. It is not practicable for the operator to hold the considerable weight of

the cylinder block, and lower it gently, whilst leading the pistons in as well. Great care has to be taken **Not** to put too much pressure on the fitted piston rings. When the pistons have finally engaged, remove the rag padding from the crankcase mouths, and the ring compressors if fitted, and lower the cylinder block still further until it seats firmly on the base gasket.

8   Take care to anchor the cam chain throughout this operation to prevent the chain dropping down into the crankcase. This task is best achieved by using a piece of stiff wire to hook the chain through, and pull up through the tunnel. The chain must engage with the crankshaft drive sprocket.

43.2 Piston must be fitted with arrow mark facing forward

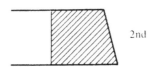

Fig. 1.14 Piston ring profiles

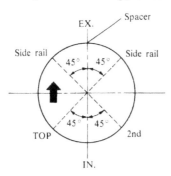

Fig. 1.15 Piston ring end gap positions

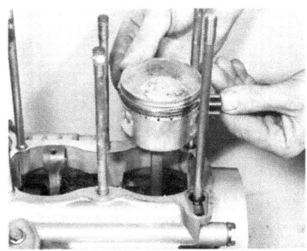

43.3 Push in gudgeon pin to secure piston

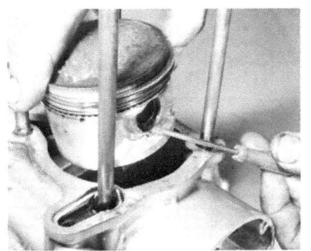

43.4 Ensure the NEW circlips are correctly fitted in groove

43.5a Refit the chain tensioner blade

43.5b Install sealing rings in the rear outer cylinder studs recesses

43.7 Guide cylinder block over pistons, two at a time; note wooden blocks for support, and use of ring clamps

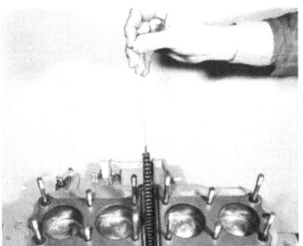

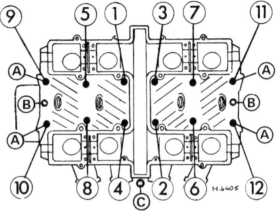

43.8 Retain cam chain by wire hook when lowering cylinder block

Fig. 1.16 Cylinder head nut torque sequence

A = Domed nuts fitted with copper washers
B = Two 6 mm bolts.
C = Third bolt fitted to late models

44 Engine reassembly: replacing the cylinder head

1   Place a new cylinder head gasket in position on the holding down studs. Fit the gasket so that the widest turned edges of the cylinder hole reinforcement inserts are facing downwards. Replace the cam tunnel rectangular seal.

2   Slide the cylinder head down the studs into position whilst guiding the cam chain through the tunnel. Secure the chain once again. Fit the cylinder head holding nuts, washers and bolts. It should be noted that the four outer studs are fitted with copper washers, to prevent oil seepage from the stud holes which serve also as oilways. Tighten the cylinder head nuts evenly in the sequence shown in the accompanying illustration, to a setting of 3·7 kgf m (26·76 lbf ft). Finally, tighten the two 6 mm bolts to 0·9 kgf m (6·50 lbf ft).

3   Insert the cam chain tensioner forward blade into the cylinder head central tunnel so that the lower end engages in the recess in the guide seat and the upper end locates with the recesses each side of the tunnel. Make a special check that the lower end of the blade is restrained; if it is allowed to float in operation the cam chain may become damaged.

44.1a Replace cam chain tunnel seal and ...

44.1b ... fit NEW cylinder head gasket

44.2a Slide the cylinder head down into position

44.2b Ensure copper washer is fitted below each domed nut

44.3 Insert forward chain guide carefully

45 Engine reassembly: replacing the camshafts and timing the valves

1   Lubricate thoroughly and insert the cam followers into their original guide tunnels in the cylinder head. Provided the followers are fitted squarely, very little pressure should be required to insert them in the tunnels. Do not attempt to tap (even lightly) a follower into place as it will probably jam solidly. Removal is then very difficult. Fit also the original adjuster shims. Lubricate the camshaft journals with a molybdenum disulphide grease and apply engine oil to the camshaft journal bearings.

2   Because the cam chain is endless, the camshaft replacement and valve timing operations must be made simultaneously. Fit a spanner to the engine turning hexagon and rotate the engine forwards until the T mark to the left of the F1-4 mark on the ATU is in **exact** alignment with the index mark on the static plate. In this position Nos. 1 and 4 cylinders are at TDC. Whilst turning the engine, the cam chain will have to be hand fed so that it does not become snagged or bunched up.

3   Select the exhaust camshaft (marked EX) and feed it through the cam chain. Pull the forward run of the chain taut and mesh the cam sprocket to the chain so that when the camshaft is lying across the bearing housings, the 2 marked arrow is pointing vertically. Insert the inlet camshaft through the cam chain in a similar way. To mesh the sprocket in the correct place, count the chain roller pins from the exhaust camshaft to

the inlet camshaft. Start with the pin directly above the 2 marked arrow on the exhaust sprocket and count to the 20th pin along the chain. Mesh the 3 marked arrow on the inlet sprocket with the 20th pin. Provided that the crankshaft has not moved during this procedure, the valve timing is now correct. As a final check that the camshafts are in the correct position, the notch on the right-hand end of each camshaft should be horizontal with the two notches facing each other.

4   The two bearing caps holding each camshaft can now be refitted. Note that each cap is marked A, B, C, or D and each should be fitted to the bearing housing similarly marked. The caps should be placed so that the letters are not inverted relative to one another. The same problem is now encountered as was found on dismantling, in that the upward pressure of the valve springs is preventing immediate seating of the camshafts. The system of clamping the camshaft down, using a self-grip wrench or G-clamp, may be adopted. Alternatively the bearing caps and bolts maybe fitted and tightened down evenly and diagonally a small amount at a time. The latter procedure should be used only if great care is exercised and neither the camshaft nor the caps are allowed to tilt. Whichever procedure is chosen the final tightening torque for the bearing cap bolts is 1·0 kgf m (7·23 lbf ft) After tightening the cap bolts, re-check the valve timing.

5   Fit the cam chain rubber bridge guide with the inscribed arrow facing forwards. Insert the tachometer driveshaft and housing into the front of the cylinder head and fit the retaining plate and bolt.

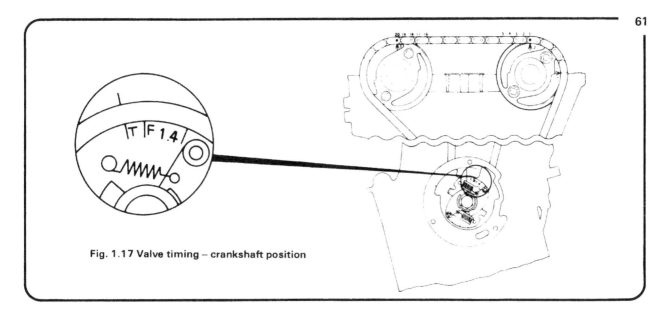

Fig. 1.17 Valve timing – crankshaft position

45.3a Each camshaft marked IN or EX (R or L) to avoid confusion

45.3b Exhaust cam sprocket timing mark (2)

45.3c Inlet cam sprocket timing mark (3)

45.3d Arrow mark (1) must be flush with cylinder head mating surface

45.4a Fit cam bearing caps so that letters match with ...

45.4b ... letters inscribed in head

45.4c Tighten the cap bolts to the correct torque figure

45.5 Refit cam chain bridge guide correctly

46 Engine reassembly: replacing and adjusting the cam chain tensioner

1   Hold the cam chain tensioner in one hand and whilst restraining the plunger rod, slacken the locking screw a few turns. Push the plunger inwards fully, simultaneously rotating the knurled adjuster wheel anti-clockwise. Continue turning until the plunger is fully retracted and the knurled adjuster has moved as far as possible. Tighten the lock screw to secure the pushrod.
2   Fit a new gasket to the tensioner body flange and install the completed unit to the rear of the cylinder block. Tighten the securing bolts evenly, to a torque setting of 0·9 – 1·4 kgf m (6·50 – 10·12 lbf ft).
3   With the tensioner fitted to the engine, unscrew the locking screw $\frac{1}{4}$ – $\frac{1}{2}$ a turn so that the plunger is free to move forwards under tension from the spring. Without allowing further rotation of the locking screw, tighten the locknut to secure the screw.
4   The cam chain tensioner is now set for automatic adjustment in service. To check that the unit is functioning correctly, rotate the engine backwards to take up all the slack in the rear run of the chain. Whilst turning the engine backwards rotate the knurled adjuster wheel slowly as far as possible. Now turn the engine in a forward direction, which will have the effect of slackening the chain on the rear run. The knurled adjuster wheel should be seen to rotate in a clockwise direction as the plunger moves out and automatically tensions the chain.
5   **WARNING.** After initial adjustment of the cam chain tensioner, the tensioner will continue to function automatically. **DO NOT** under any circumstances rotate the knurled adjuster wheel either clockwise or anti-clockwise except when making this adjustment in the prescribed manner. Rotation of the wheel except at this stage will cause excessive chain tightness and will lead to tensioner damage and chain damage.

47 Engine reassembly: checking and adjusting the valve clearances

1   The clearance between each cam and cam follower must be checked and if necessary adjusted by removal of the existing adjuster pad and replacement of a pad of suitable thickness. Make the clearance check and adjustment of each valve in sequence and then go on to the next valve. As shown in the diagram in the Routine Maintenance section both operations should be carried out with the cam lobe in question placed in one of two alternative positions.

2   Using a feeler gauge, verify and record the clearance at the first valve. If the clearance is incorrect, not being within the range 0·03 – 0·08 mm (0·001 – 0·003 inch), the adjuster pad must be removed and replaced by a shim of suitable thickness. A special tool is available (Suzuki part no. 09916-64510) which may be pushed between the camshaft adjacent to the cam lobe and the raised edge of the cam follower, to allow removal of the shim. If the special tool is not available, a simple substitute may be fabricated from steel plate. The final form of the tool which has a handle approximately 6 inches long is shown in the photograph accompanying the cam adjustment section in the Routine Maintenance Chapter. The Suzuki tool may be pushed into position, depressing and securing the cam follower in a depressed position in one operation. Where a home-made tool is used, the cam follower may be depressed by inserting a suitable lever between the adjuster pad and the cam lobe. The tool may then be inserted to secure the cam follower whilst the adjuster pad is removed. Before installing either type of tool, rotate the cam follower so that the slot in the follower raised edge is not observed by the camshaft. Insert a screwdriver through the slot to displace the adjuster pad.

3   Adjustment pads are available in 20 sizes ranging from 2·15 mm to 3·10 mm in increments of 0·05 mm. Each pad is identified by a three digit number etched on the reverse face. The number (eg 235) indicates that the pad thickness is 2·35 mm thick. To select the correct pad subtract 0·03 mm from the measured clearance and add the resultant figure to that of the existing pad. Select the largest available pad whose thickness is the same as, or failing that, slightly smaller than the final figure. Refer to the table accompany the Routine Maintenance Section for the selection of available pads.

Although the adjustment pads are available as a set, their price is prohibitive. It is suggested that pads are purchased individually, after an accurate assessment of requirements has been made. It is possible that some Suzuki Service Agents will be prepared to exchange used pads for others of the correct size, providing that the original pads are not worn.

4   Before inserting a replacement pad, lubricate both sides thoroughly with engine oil. Always fit the pads with the number downwards, so that it does not become obliterated by the action of the cam. Recheck the valve clearance after fitting the new adjuster pads.

5   After restoring the clearances on each valve refit the camshaft cover. Do not omit the semi-circular seals which are located at each end of the camshaft chambers. Replace the camshaft cover end caps.

46.2 Fit chain tensioner, using a new gasket

47.5a After fitting a new gasket ...

47.5b ... and checking the condition of cam chain guide

47.5c ... refit the camshaft cover

48 Engine reassembly: replacing the engine in the frame

1 The task of replacing the engine requires three people, two to lift the engine and one to hold the frame steady while the engine is lowered into position.

2 Lift the complete engine unit into the frame from the right-hand side and mount but do not secure the four engine mounting brackets, before inserting the engine bolts. Insert the three long bolts from the left-hand side of the machine and fit the two short central bolts and the special plate bolts. Do not force the bolts into position as this may damage the threads. Use a wooden lever between the frame and the casings, in order to lift the engine and align bolt holes. Tighten the engine mounting bracket bolts first and then the engine bolts.

3 Install the final drive sprocket to the engine with the chain already fitted to the sprocket, tilt the lockwasher and tighten the locknut. Bend up the lockwasher to secure the nut.

4 Replace the starter motor in the compartment and fit and tighten the screws. Reconnect the starter motor lead with the terminal on the motor body. Remake the connections from the alternator, contact breaker, neutral indicator or gear indicator switch and oil pressure warning lamp leads. Ensure that the leads are routed correctly and secured as necessary. Ensure that the colour coding of the wires is followed exactly. Reconnect the main earth lead to the crankcase bolt.

5 Replace the final drive sprocket cover. Reconnect the clutch cable and adjust the cable so that there is 4mm ($\frac{1}{8}$ in) play measured between the stock and lever, before the clutch commences lifting.

6 Replace the breather cover on the cam box cover together with a new gasket.

7 Position the air filter box to the rear of the engine and then lift the carburettor into place from the right-hand side. Ensure that the screw clips are fitted to the induction stubs and air hoses before installing the carburettors. Fit the air filter box mounting bolts and then tighten the screw clips fully.

Refit the two throttle cables to the wheel type control lever (one cable opens the throttles, and one cable closes them), making sure that the opening cable is fitted to the rear, and the closing cable is fitted to the front of the operating wheel.

8 Install the connecting hose between the breather cover and the airfilter box. Secure the hose by means of the spring clips.

The carburettor drain hoses should be gathered together and retained by a strap to the engine right-hand upper rear mounting bracket. Allow the trailing ends of the hoses to pass between the swinging arm member and the rear of the engine. The air vent pipes from carburettors No. 2 and No. 4 should be passed through the clamp attached to the front of the rear air box.

9 Refit the exhaust pipes and also the silencers if they too were removed. Replace the gearchange lever, checking it is in the correct operating position before tightening the pinch bolt.

10 Replace the rider's footrests, tightening fully the two securing bolts on each.

11 If the coils were detached during removal of the engine from the frame stage, they should now be replaced along with their HT leads. Reconnect the HT leads with the sparking plugs. Each cable is marked by a numbered sleeve to aid correct fitting.

12 If the two horns were removed previously, they must now be replaced. When refitting, ensure that they are positioned correctly and will not come into contact with the petrol tank during the running of the machine.

13 Lower the petrol tank into place and refit the rear retaining bolt and rubber seat. Connect the petrol pipe and the vacuum tube to the petrol tap unions. Secure the petrol pipe by means of the spring clip and reconnect the two fuel gauge leads at their snap connectors.

14 Check that the crankcase drain plug has been secured, and then refill the engine with the correct amount of engine oil. The leveL can be checked through the sight window in the clutch cover which should be between the two marks. Replenish, using the specified quantity of SAE 10W/40 engine oil.
Oil capacity:
　　3·4 ltr (7·2/6·4 US/Imp pints
Allow the engine to run for approximately 3 minutes after the initial start-up and then recheck the oil level. Ensure that the machine is standing vertically when checking the level because any angle of lean has a marked effect on the indicated level.

15 Replace the battery in its carrier below the air filter and remake the connections, Give both terminals a coat of petroleum jelly to inhibit corrosion. Refit the rear air filter box and air filter contained inside. Ensure the clip on the hose joint between the front and rear air filter boxes is tightened correctly. Fit the two frame side covers.

48.3a Mesh sprocket with chain and fit on shaft

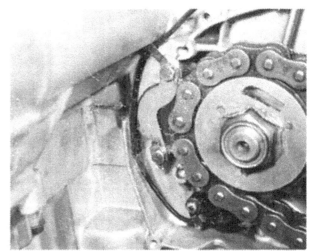

48.3b After tightening centre nut bend up tab washer

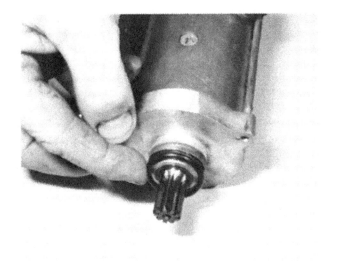

48.4a Replace starter motor ensuring O ring is in good condition

48.4b Reconnect the starter motor lead with terminal and ...

48.4c ... refit chromed motor cover

48.4d Reconnect oil pressure warning lamp switch lead

48.6 Replace cam box breather cover

48.7a With air filter box installed, refit the carburettors

48.7b Refit the two throttle cables and breather hose

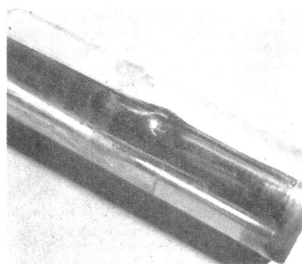

48.9a Exhaust outer pipes marked (R or L) to aid reassembly

48.9b Note fitting of split collars to centre pipes only

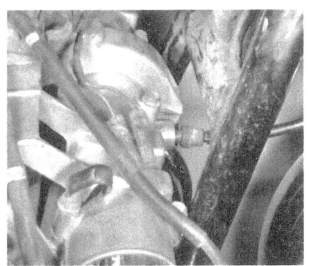

48.12 Reconnect both horn leads and replace HT leads in the retaining clips

48.14 Check oil level through sight window in clutch cover

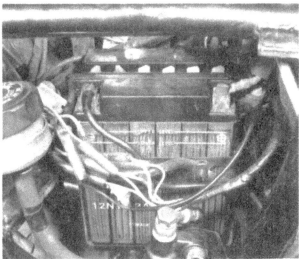

48.15 Replace battery in its carrier below air filter case

49 Starting and running the rebuilt engine

1   Open the petrol tap, close the carburettor chokes and start the engine. Raise the chokes as soon as the engine will run evenly and keep it running at a low speed for a few minutes to allow oil pressure to build up and the oil to circulate. If the red oil pressure indicator lamp is not extinguished, stop the engine immediately and investigate the lack of oil pressure.

2   The engine may tend to smoke through the exhausts initially, due to the amount of oil used when assembling the various components. The excess of oil should gradually burn away as the engine settles down.

3   Check the exterior of the machine for oil leaks or blowing gaskets. Make sure that each gear engages correctly and that all the controls function effectively, particularly the brakes. This is an essential last check before taking the machine on the road.

50 Taking the rebuilt machine on the road

1   Any rebuilt machine will need time to settle down, even if parts have been replaced in their original order. For this reason it is advisable to treat the machine gently for the first few miles to ensure oil has circulated throughout the lubrication system and that any new parts fitted have begun to bed down.

2   Even greater care is necessary if the engine has been rebored or if a new crankshaft has been fitted. In the case of a rebore, the engine will have to be run-in again, as if the machine were new. This means greater use of the gearbox and a restraining hand on the throttle unit until at least 500 miles have been covered. There is no point in keeping to any set speed limit, the main requirement is to keep a light loading on the engine and to gradually work up performance until the 500 mile mark is reached. These recommendations can be lessened to an extent when only a new crakshaft is fitted. Experience is the best guide since it is easy to tell when an engine is running freely.

3   If at any time lubrication failure is suspected, stop the engine immediately, and investigate the cause. If an engine is run without oil, even for a short period, irreparable engine damage is inevitable.

51 Fault diagnosis: engine

Symptom	Cause	Remedy
Engine will not start	Defective sparking plugs	Remove the plugs and lay them on the cylinder head. Check whether spark occurs when ignition is on and the engine is rotated.
	Dirty or closed contact breaker points	Check the condition of the points and whether the points gaps are correct.
	Faulty or disconnected condenser	Check whether the points arc when separated. Renew the condenser if there is evidence of arcing.
Engine runs unevenly	Ignition and/or fuel system fault	Check each system individually, as though engine will not start.
	Blowing cylinder head gasket	Leak should be evident from oil leakage where gas escapes. High pitched squeak may be evident with engine running.
	Incorrect ignition timing	Check accuracy and, if necessary, reset.
Lack of power	Fault in fuel system or incorrect ignition timing	See above
Heavy oil consumption	Cylinder block in need of rebore	Check bore wear, rebore and fit over-size pistons if required.
	Damaged oil seals	Check engine for oil leaks.
Excessive mechanical noise	Worn cylinder bores (piston slap)	Rebore and fit oversize pistons.
	Worn camshaft drive chain (rattle)	Adjust tensioner or replace chain.
	Worn big-end bearings (knock)	Fit replacement crankshaft assembly.
	Worn main bearings (rumble)	Renew crankshaft assembly.
Engine overheats and fades	Lubrication failure	Stop engine and check whether internal parts are receiving oil. Check oil level in crankcase.

52 Fault diagnosis: clutch

Symptom	Cause	Remedy
Engine speed increases as show by tachometer, but machine does not respond	Clutch slip	Check clutch adjustment for free play at handlebar lever. Check thickness of inserted plates.
Difficulty in engaging gears; gear changes jerky and machine creeps forward when clutch is withdrawn, difficulty in selecting neutral	Clutch drag	Check clutch adjustment, at actuating lever on crankcase for too much free play. Check clutch drum for indentations and check clutch plates for burrs on tongues. Dress with file if damage not too great.
Clutch operation stiff	Damaged, trapped, or frayed control cable	Check cable and replace if necessary. Make sure cable is lubricated and has no sharp bends.

53 Fault diagnosis: gearbox

Symptom	Cause	Remedy
Difficulty in engaging gears	Selector forks bent	Replace with new forks
	Gear clusters not assembled correctly	Check gear clusters for arrangement and position of thrust washers
Machine jumps out of gear	Worn dogs on ends of gear pinions	Renew worn pinions
Gear change lever does not return to original position	Broken return spring	Renew spring

Chapter 2 Fuel system and lubrication

Contents

Specifications

Fuel tank capacity

Overall	19 lit (5.0/4.1 US/Imp gal)
Reserve	4.0 lit (8.4/7.0 US/Imp pint)
GS1000 L:	
Overall	15 lit (4.0/3.3 US/Imp gal)
Reserve	3.0 lit (6.4/5.3 US/Imp pint)

Carburettors

Make	Mikuni
Type	VM26SS (VM28SS)*
Main jet	95
Main air jet	1.5
Needle jet	AO – 2
Jet needle	5DL36-3
Pilot jet	15
Pilot air jet	1.2
Throttle valve cutaway	1.5
Air screw	Pre-set (**Do not** disturb)
Pilot screw	Pre-set (**Do not** disturb)
Float height	23 – 25 mm (0.905 – 0.984 in)

*VM28SS fitted to UK spec. GS1000 S and EN models.

Engine/transmission oil

Capacity:	
Dry	4.2 lit (8.8/7.4 US/Imp pint)
Without filter change	3.4 lit (7.2/6.0 US/Imp pint)
With filter change	3.8 lit (8.0/6.6 US/Imp pint)
Specification	SAE 10W/40

Oil pump

Type	Trochoid
Output pressure	0.1 kg/cm² (1.42 psi)
Inner rotor/outer rotor clearance	0.12 mm (0.004 in)
Service limit	0.2 (0.007 in)
Outer rotor/housing clearance	0.07 – 0.160 mm (0.002 – 0.006 in)
Service limit	0.25 mm (0.009 in)
Side clearance	0.04 – 0.09 mm (0.001 – 0.003 in)
Service limit	0.15 mm (0.006 in)

1 General description

The fuel system comprises a petrol tank from which petrol is fed by gravity to the float chamber of each of the four carburettors. A single petrol tap with a detachable gauze filter is located beneath the petrol tank, on the left-hand side. It contains provision for a reserve quantity of petrol, when the main supply is exhausted. The petrol tap has three positions; ON, RES, and PRI. Fuel can, however, only flow to the carburettors when the engine is running, with the top in ON or RES position. This is due to the tap diaphragm which is controlled by the induction pressure. If there is no fuel in the float chambers, as may be the case after carburettor dismantling, the petrol tap should be turned to the PRIMING position, to allow an unrestricted flow of petrol to the float chambers. Return the tap to the ON position as soon as the engine is running.

Four Mikuni throttle slide carburettors are fitted of VM26SS type to all models except the 1000 'S' and EN models in the UK which are fitted with VM28SS instruments. The carburettors are mounted as a unit on a cast aluminium alloy bracket. They are controlled by a push-pull cable arrangement secured to a cross-rod, which passes through the top of each carburettor, connecting each throttle valve by a bell-crank arrangement.

For cold starting, a hand-operated choke lever attached to the far left-hand carburettor is linked to the three other carburettors, so that the mixture can be enriched temporarily. When the engine has started, the choke can be opened gradually as the engine warms up, until full air is accepted under normal running conditions.

Lubrication is effected by the wet sump principle in which the reservoir of oil is contained within engine sump. This oil is shared by the engine, primary drive and transmission components. The oil pump is of the Eaton trochoid type and is driven from a pinion engaged with and to the rear of the clutch.

Oil is supplied under pressure, via a full flow oil filter with a replaceable element to the crankshaft and to the overhead camshaft and rocker gear. A secondary flow passes to the gearbox via the gearbox main bearings. All surplus oil drains to the sump and is returned to the oil tank by the scavenge section of the oil pump. The pump itself is protected by a gauze strainer in the base of the oil pick-up chamber.

2 Petrol tank: removal and replacement

1   The petrol tank is retained by two guide channels which locate with a circular rubber block on each side of the steering head and a bolt passing through a lug at the rear of the tank and into the frame.

2   To remove the tank, leave the petrol tap lever at the ON or RESERVE position and detach the diaphragm vacuum pipe and the larger bore petrol feed pipe. The latter pipe is secured at the union by a spring clip, the ears of which should be pinched together to release the tension on the pipe. Remove the bolt from the rear of the tank, after raising the dual seat to gain access.

3   Detach the lead for the fuel gauge at the snap connector below the left-hand side of the petrol tank.

4   Lift the tank up at the rear and then ease the unit backwards until the location cups leave the rubber blocks. At this stage the tank is obstructed by the location cups coming in contact with the ignition coils bolted to the frame. Lift the tank upwards off the machine.

5   Drainage of petrol is not strictly necessary when removing the tank, although the reduction in weight will facilitate the operation. A full tank will weigh approximately 50 lbs.

6   Replace the petrol tank by reversing the removal procedure. Take care not to trap control cables or stray wires between the tank and frame tubes. If the cups are a tight fit on the rubber blocks, apply a small amount of washing-up detergent to the blocks to ease refitting.

3 Petrol tap: removal and replacement

1   Removal of the complete petrol tap is required at regular intervals to gain access to the filter columns for cleaning. The tank should be drained of petrol by fitting a length of tubing to the petrol tap outlet and turning the lever to the Priming position.

2   The petrol tap is held to the underside of the petrol tank by two crosshead screws with washers. Note that there is an O-ring seal between the petrol tap body and the petrol tank, which must be renewed if it is damaged, or if petrol leakage has occurred. The filter screens which are integral with the plastic level pipes should be cleaned of any deposits using a soft brush and clean petrol. Because there is only a single tap to feed four carburettors, any restriction in petrol flow may lead to fuel starvation, causing missing and in extreme cases overheating, due to a weak mixture.

3   It is seldom necessary to remove the lever which operates the petrol tap, although occasions may occur when a leakage develops at the joint. Although the tank must be drained before the lever assembly can be removed, there is no need to disturb the body of the tap.

2.1 Petrol tank is retained by rubber blocks at forward end

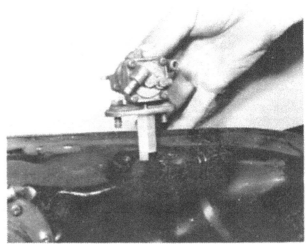

3.2 Petrol tap held to tank underside by two bolts

4   To dismantle the lever assembly, remove the two crosshead screws passing through the plate on which the operating positions are inscribed. The plate can then be lifted away, followed by a spring, the lever itself and the seal behind the lever. The seal will have to be renewed if leakage has occurred. Reassemble the tap in the reverse order. Gasket cement or any other sealing medium is NOT necessary to secure a petrol tight seal.

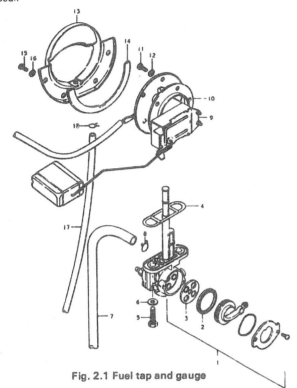

Fig. 2.1 Fuel tap and gauge

1	Fuel tap assembly	10	Sealing ring
2	Tap lever gasket	11	Bolt – 5 off
3	Tap valve gasket	12	Spring washer – 5 off
4	O-ring	13	Fuel drain plate
5	Bolt – 2 off	14	Gasket
6	Sealing washer – 2 off	15	Screw – 4 off
7	Fuel delivery pipe	16	Spring washer – 4 off
8	Clip – 2 off	17	Overflow pipe
9	Fuel level gauge assembly	18	Clip

4   Carburettors: removal from the machine

1   To improve access to the carburettors it is suggested that the petrol tank is removed, as described in Section 2 of this Chapter, before dismantling proper commences.

2   Before detaching the air filter box or carburettors displace the various vent and drain tubes from the securing clips.

3   Detach the engine breather hose from the unions at the air filter box and the breather cover on the cylinder head. The hose is secured at both ends by spring clips. Disconnect the throttle cables from the operating pulley at the carburettors. Both may be detached in a similar manner. Loosen the upper and lower locknuts on the cable adjuster screw and displace the adjuster and outer cable from the abutment bracket. Rotate the pulley until the inner cable nipple can be pushed out of the anchor point.

4   Loosen the screw clips which secure the air filter hoses and inlet stubs to the carburettors. Remove the air filter mounting bolts and ease the box rearwards, so that the hoses leave the carburettor mouths.

5   The carburettors must be removed without first removing the front air filter box. Problems will be encountered if the operator attempts to remove the front air box before the carburettors, as the latter component is obstructed by the frame tubes.

5   Carburettors: dismantling and reassembly

1   The carburettors are mounted on a cast aluminium bracket which also serves as a support for the choke operating link rod and the cable anchor bracket. The bracket is so arranged that partial dismantling of all the carburettors is required before they can be removed from the bracket and attended to as individual items. Whenever practically possible, dismantle the carburettors separately. This lessens the chance of accidentally mixing up the various components of individual carburettors.

2   Remove the tops from the four carburettors. Each top is retained by three screws. Remove the bolt which passes through the forward end of each bellcrank and locates with the throttle link shaft. The throttle shaft is located longitudinally by a claw plate, secured by a single screw to a lug projecting from the mounting brackets. Remove the claw plate and then prise out the outer blind grommets from the left-hand and right-hand carburettors. Using a pair of snipe nose pliers, disconnect the throttle pulley return spring from the two anchor pegs. The shaft can now be pushed out of position to free the bell cranks and the pulley wheel. Note that the pulley cannot be removed completely until the carburettors are separated.

3   Place the carburettors rear face downwards and remove the two countersunk screws which hold each instrument to the mounting bracket. Lift the bracket away, if necessary moving the choke link rod slightly so that the operating arm forks clear the choke plungers.

4   The carburettors are now joined only by the fuel cross feed pipes, which are a push fit in the bodies. Before separating the individual units, mark each carburettor carefully so that on reassembly no confusion arises as to their correct positions.

5   Select one carburettor and continue dismantling as follows, following suit with the other three carburettors, in turn. Lift out the bell crank, link arm and throttle valve unit, taking care not to damage the throttle needle. The throttle valve and needle may be removed from the valve seat by removing the two tiny screws. Invert the valve and allow the needle and needle clip to fall out. Removal of the link arm from the bell crank requires that the throttle slide vertical position adjuster screw is unscrewed fully and hence the original adjustment will be lost. It is unlikely that the link arm and bell crank will require separation and it is therefore advised that these components are left as a sub-assembly.

6   Invert the carburettor and remove the four screws that hold the float chamber to its base. Remove the hinge pin that locates the twin float assembly in each carburettor, and lift away the float. This will expose the float needle. The needle is very small and should be put in a safe place so that it is not misplaced.

7   Make sure the float chamber gasket is in good condition. It should not be disturbed unless it shows sign of damage or has been leaking.

8   Unscrew the main jet from the jet holder, using a wide bladed screwdriver and then unscrew the holder itself. Note the O-ring fitted above the holder threads. Invert the carburettor body and displace the needle jet from the central bore.

9   Note that the pilot mixture adjusting screw and the pilot air adjusting screw **Must Not** be removed or adjusted. Both of these components are pre-set under specialised factory conditions and cannot be re-set, when once maladjusted, except under the same conditions. Further to this, all GS1000 models produced after January 1st, 1978, are equipped with specially manufactured carburettors, with certain components machined to extremely close tolerances. There are three specific components, in each group of mixture control components, that are of a particularly close tolerance; the main jet, the needle jet and the pilot air jet. If replacement of any of these jets becomes

necessary, a new part of the same, close tolerance type, must be obtained. To aid the operator in selecting the correct jet, the three later type jets use a different style of numerical identification. See accompanying table. The reasons for the adoption of pre-set mixture adjustments and extremely close tolerance carburettor jetting, is ostensibly to enable the GS1000 to meet the US Federal Emission Regulations. Therefore, by adjusting, resetting, or replacing the pilot air or pilot mixture screw, the emission levels may be adversely affected, thus leading to a possible infringement of State or Federal emission regulations. This could subsequently render the operator subject to a fine. It is, therefore, strongly recommended that the advice of your local Suzuki dealer be sought, before attempting any operation controlling the jetting and/or the mixture of the carburettors.

Conventional Figures Used On Standard Tolerance Jet Components	1 2 3 4 5 6 7 8 9 0
Emission Type Figures Used On Close Tolerance Jet Components	1 2 3 4 5 6 7 8 9 0

10 The yellow-painted fuel metering screws **Must Not** be disturbed. These, too, are set at the factory to suit individual carburettors.

11 The starter plunger (choke) assembly is positioned in a tunnel to the side of the upper chamber. Unscrew the housing cap and pull the starter plunger assembly out. This consists of the plunger rod, spring and plunger piece.

12 Check the condition of the floats. If they are damaged in any way, they should be renewed. The float needle and needle seating will wear after lengthy service and should be inspected carefully. Wear usually takes the form of a ridge or groove, which will cause the float needle to seat imperfectly. Always

renew the seating and needle as a pair. An imperfection in one component will soon produce similar wear in the other.

13 After considerable service the throttle needle and the needle jet in which it slides will wear, resulting in an increase in petrol consumption. Wear is caused by the passage of petrol and the two components rubbing together. It is advisable to renew the jet periodically in conjunction with the throttle needle.

14 Before the carburettors are reassembled, using the reversed dismantling procedure, each should be cleaned out thoroughly using compressed air. Avoid using a piece of rag since there is always risk of particles of lint obstructing the internal passageways or the jet orifices.

15 Never use a piece of wire or any pointed metal object to clear a blocked jet. It is only too easy to enlarge the jet under these circumstances and increase the rate of petrol consumption. If compressed air is not available, a blast of air from a tyre pump will usually suffice.

16 Do not use excessive force when reassembling a carburettor because it is easy to shear a jet or some of the smaller screws. Furthermore, the carburettors are cast in a zinc-based alloy which itself does not have a high tensile strength. Take particular care when replacing the throttle valves to ensure the needles align with the jet seats.

17 Reassemble the carburettors by reversing the dismantling procedure. Before inserting the throttle link shaft, it should be lubricated with grease. Check the condition of the two O-rings which seal each side of the fuel transfer pipes. Renew them if there is any doubt as to their efficiency. Fit and tighten the eight carburettor mounting screws before tightening fully the through bolts which secure the bell cranks to the throttle shaft. A small amount of locking fluid should be applied to the mounting screw threads before they are inserted.

18 Before replacing the carburettors on the machine and before refitting the carburettor tops, refer to the next section for details of carburettor synchronisation.

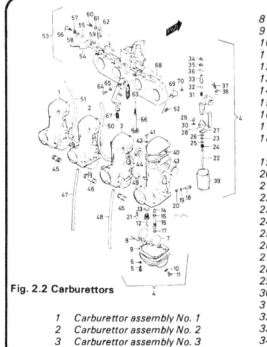

Fig. 2.2 Carburettors

1   Carburettor assembly No. 1
2   Carburettor assembly No. 2
3   Carburettor assembly No. 3
4   Carburettor assembly No. 4
5   Screw – 16 off
6   Spring washer – 16 off
7   Float assembly – 4 off
8   Pivot pin – 4 off
9   Float chamber gasket – 4 off
10 O-ring – 4 off
11 Screw – 4 off
12 Float needle valve assembly – 4 off
13 Needle valve gasket – 4 off
14 Needle jet – 4 off
15 Jet holder – 4 off
16 O-ring – 4 off
17 Main jet – 4 off
18 Mixture adjustment (pilot air) screw – 4 off
19 O-ring – 4 off
20 Spring – 4 off
21 Pilot jet – 4 off
22 Jet needle – 4 off
23 Needle clip – 4 off
24 Spring – 4 off
25 Spring seat – 4 off
26 Ring – 4 off
27 Needle seat – 4 off
28 Throttle valve linkage – 4 off
29 Screw – 8 off
30 Spring washer – 8 off
31 Spring – 4 off
32 Seat – 4 off
33 Adjusting screw – 4 off
34 Nut – 4 off
35 Spring washer – 4 off
36 Washer – 4 off
37 Bolt – 4 off
38 Spring washer – 4 off
39 Throttle valve – 4 off
40 Gasket – 4 off
41 Screw – 12 off
42 Spring washer – 12 off
43 Blind grommet – 2 off
44 Grommet – 6 off
45 Union – 2 off
46 Union
47 Overflow pipe – 2 off
48 Overflow pipe – 2 off
49 Fuel transfer pipe – 2 off
50 Fuel delivery pipe
51 Fuel delivery pipe – 2 off
52 Screw – 8 off
53 Choke lever assembly
54 Choke control shaft
55 Washer
56 Screw
57 Spring
58 Washer
59 Washer
60 Screw
61 Spring washer
62 Bracket
63 Throttle adjuster return spring
64 Screw – 2 off
65 Spring washer – 2 off
66 Remote throttle adjuster
67 Spring
68 Throttle link rod
69 Choke operating fork – 4 off
70 Screw – 4 off

5.2a Remove bolt from each bellcrank and ...

5.2b ... remove the screw retaining the claw plate and ...

5.2c ... detach the claw plate which secures the throttle rod

5.2d Prise out the blind end plugs and ...

5.2e ... disconnect the throttle pulley return spring and ...

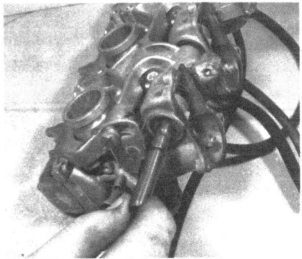

5.2f ... withdraw the throttle rod

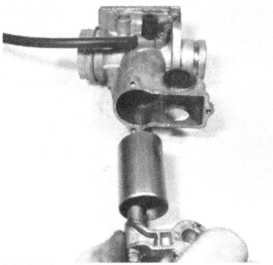

5.5a Withdraw the throttle valve unit

5.5b Crank link rod is secured by two tiny screws

5.6a Float chamber is held by four screws

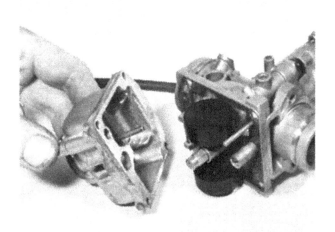

5.6b Removal of the screws allows float chamber to be separated

5.6c Push out pivot pin to detach float assembly

5.6d Do not misplace the small float needle (arrowed); its seat may be removed for renewal

5.7 Check that the float chamber gasket is in good condition

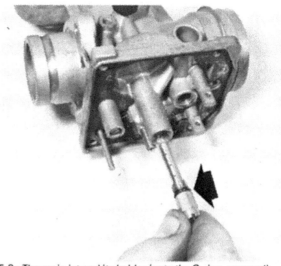

5.8a The main jet and its holder (note the O-ring, arrowed)

5.8b Displace the needle jet from the central bore

5.11 Choke plunger rarely gives trouble

5.17a Check condition of fuel transfer pipe O-rings

5.17b Refit and tighten float chamber drain plug, if it was removed

5.18 Ensure carburettors are readjusted before refitting the carburettor tops

6   Carburettors: synchronisation and adjustment

1   Synchronisation of the carburettors should be carried out in two stages. The first stage as detailed in the following paragraph should be accomplished with the carburettors removed from the machine, at any time after the carburettors have been dismantled and reassembled. The second stage, which is synchronisation of the carburettors using vacuum gauges, should be carried out as a routine maintenance item. It is also necessary when rough idling or poor engine performance is encountered, or after the carburettors have been refitted to the machine.
2   Unscrew the remote throttle stop screw which is fitted with a nylon head, so that clearance can be seen between the end of the screw and the portion of the throttle pulley against which it normally abuts. The throttle slides should now be in the fully closed position. Make a visual check that all four throttle slides open and close at precisely the same time. If variations occur, each slide may be adjusted individually by means of the adjuster screw on the end of the bell crank. Loosen the locknut on the screw and make the required adjustment. Retighten the locknut without moving the screw. Rotate the throttle pulley so that the slides are in the fully open position and check that the lower edge of the slides are at the position indicated in the accompanying diagram. Adjustment of this setting should be

made by turning the single, spring secured screw, which is fitted to the rear face of the throttle cable anchor plate. After making the fully open and fully closed adjustments, replace the carburettor tops and refit the carburettors and controls to the machine.
3   Adjust the throttle cables by starting with the opening cable first. Loosen the locknut on the throttle opening cable, and use the adjuster to allow 1.0–1.5 mm (0.04–0.06 in) of slack in the cable before securing the locknut again. Loosen the locknut on the closing cable, and adjust it so there is no play in the throttle grip, then secure the locknut.
4   As stated above, running adjustment of the carburettors requires the use of a set of vacuum gauges or indicators, together with the appropriate adaptors, which screw into the inlet tracts and to which are attached the vacuum pipes. Unless the vacuum gauge set is to hand, it is recommended that the machine is returned to a Suzuki Service Agent who will carry out the synchronisation as a normal service task.
5   Remove the blanking screws from the inlet tracts on the cylinder head, fit the adaptors and connect up the vacuum gauges. Start the engine and allow it to run until normal running temperature has been reached. By means of the throttle stop screw raise the engine speed to a steady 1,500 rpm. Select as a datum the carburettor which shows the central reading. Make adjustment using the bell-crank adjuster screw in each carburettor so that the three readings on the remaining gauges are modified to correspond with that of the datum gauge. Using the throttle stop screw reduce the engine speed to the specified tick-over speed of 1,000 rpm. Disconnect the vacuum gauges and refit the take-off blanking plugs, together with their sealing washers.

7   Carburettors: checking the float chamber fuel level

1   If conditions of a continual weak mixture or flooding are encountered on one or more carburettors, or if difficulty is experienced in tuning the carburettors, the float levels should be checked and, if necessary, adjusted. Although the float chambers may be removed with the carburettors in situ on the machine, it is advised that the carburettors be removed to facilitate inspection and adjustment.
2   The float level is correct when the distance between the uppermost edge of the floats (with the carburettor inverted) and the mixing chamber body flange is 23-25 mm (0.905-0.984 in). The gasket must be removed from the mixing chamber body before the measurement is taken. The floats should be in the closed position when the measurement is taken. Adjustment is made by bending the float assembly tang (tongue), which engages with the float, in the direction required.

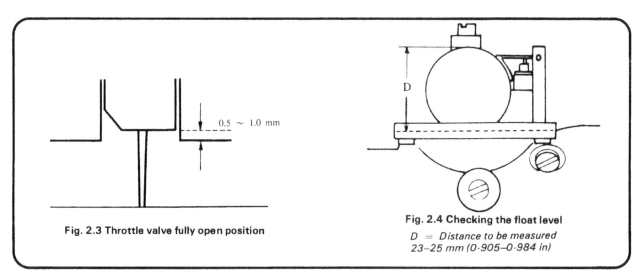

Fig. 2.3 Throttle valve fully open position

Fig. 2.4 Checking the float level

D = Distance to be measured
23–25 mm (0·905–0·984 in)

8 Carburettors: settings

1   As stated in Section 5, paragraph 9 of this Chapter, most of
the carburettor settings, such as the size of the needle jets, main
jets and needle positions, are pre-determined by the
manufacturer. It was also noted that due to current and future
emissions regulations, the amount of choice over these component settings is now non-existent. It is, however, unlikely that
under normal riding conditions, these settings will require
modification. If a change does appear necessary, it can often be
traced to an engine fault, or an engine modification, eg; a leaky
exhaust port joint or an after-market exhaust system.
2   As an appropriate guide to the carburettor settings, the pilot
jet controls the engine speed up to 1/8th throttle. The throttle
slide cut-away controls the engine speed from 1/8th to 1/4th
throttle and the position of the needle in the slide from 1/4 to
3/4 throttle. The size of the main jet is responsible for engine
speed at the final phase of 3/4 to full throttle. These are only
guide lines; there is no clearly defined demarcation line due to a
certain amount of overlap that occurs.
3   Always err slightly towards a rich mixture as one that is too
weak will cause the engine to overheat and burn the exhaust
valves. Reference to Chapter 3 will show how the condition of
the sparking plugs can be interpreted with some experience as a
reliable guide to carburettor mixture strength.
4   Alterations to the mid-range mixture strength can be made

by changing the position of the throttle needle in the throttle
slide by moving the needle clip into a different groove. Raising
the needle will richen the mixture and lowering the needle will
weaken it.

9 Air cleaner: dismantling, servicing and reassembly

1   The air cleaner on the GS1000 models is divided into a
front and rear box. The forward air cleaner box is mounted
immediately behind the four carburettors into which the carburettor intakes fit. The rearward air filter box contains the
removable filter element. The rear, element-containing box, is
situated below the dualseat, between the frame tubes. The filter
element, of the dry paper type, is removable for cleaning or
replacement, whichever is appropriate.
2   To gain access to the filter element, first raise the dualseat.
The element box lid pivots at the forward end, and is secured by
one screw at the other end. Remove the screw, displace the lid,
and pull out the element by grasping the sprung retainer at the
forward end.
3   The element should be cleaned every 2000 miles (3000
kms) and replaced every 8000 miles (12000 kms)
4   Never run the machine without the element or with the air
cleaner disconnected, otherwise the weak mixture that results
will cause engine overheating and severe damage.

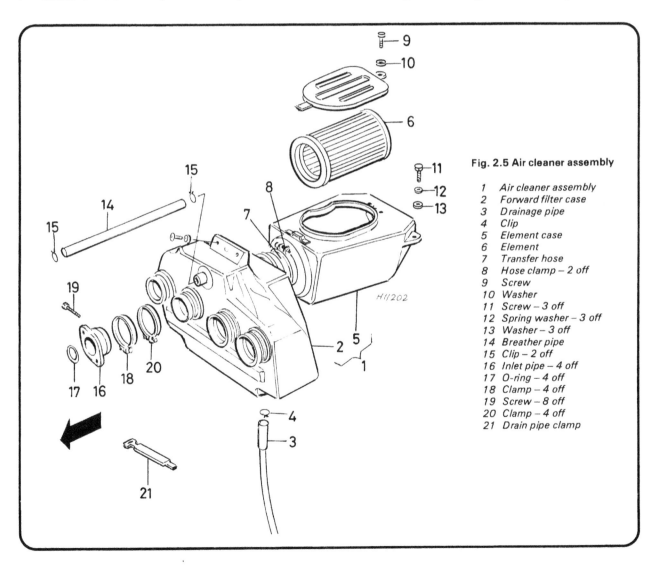

Fig. 2.5 Air cleaner assembly

1   Air cleaner assembly
2   Forward filter case
3   Drainage pipe
4   Clip
5   Element case
6   Element
7   Transfer hose
8   Hose clamp – 2 off
9   Screw
10 Washer
11 Screw – 3 off
12 Spring washer – 3 off
13 Washer – 3 off
14 Breather pipe
15 Clip – 2 off
16 Inlet pipe – 4 off
17 O-ring – 4 off
18 Clamp – 4 off
19 Screw – 8 off
20 Clamp – 4 off
21 Drain pipe clamp

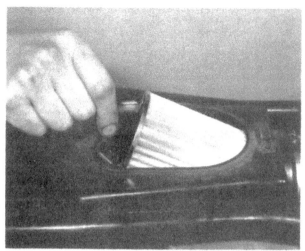

9.2 Filter element pulls out of spring retainer

10 Engine and gearbox lubrication

1   As previously described at the beginning of the Chapter the lubrication system is of the wet sump type, with the oil being forceably pumped from the sump to positions at the gearbox bearings, the main bearings, and the cam box bearings, all oil eventually draining back to the sump. The system incorporates a gear driven oil pump, an oil filter, a safety by-pass valve, and an oil pressure switch. Oil vapours created in the crankcase are vented through a breather to the air cleaner box, where they are passed into the cylinder providing an oil-tight system.

2   The oil pump is an Eaton trochoid twin rotor unit which is driven from a gear engaged with and to the rear of the clutch. An oil strainer is fitted to the intake side of the pump, which serves to protect the pump mechanism from impurities in the oil which might cause damage.

3   A corrugated paper oil filter is included in the system and is fitted within an enclosed chamber in the front of the crankcase. Access to the filter is made through a finned cover. As the oil filter unit becomes clogged with impurities, its ability to function correctly is reduced, and if it becomes so clogged that it begins to impede the oil flow, a by-pass valve opens, and routes the oil flow through the filter core. This results in unfiltered oil being circulated throughout the engine, a condition which is avoided if the filter element is changed at regular intervals.

4   The oil pressure switch, which is situated at the top of the crankcase behind the cylinder block, serves to indicate when the oil pressure has dropped due to an oil pump malfunction, blockage in an oil passage, or a low oil content. The switch is not intended to be used as an indication of the correct oil level.

5   As previously mentioned an oil breather is incorporated into the system. It is mounted in the top of the camshaft cover and is essential for an engine of this size with so many moving parts. It serves to minimise crankcase pressure variations due to piston and crankshaft movement, and also helps lower the oil temperature, by venting the crankcase. Furthermore this system reduces the escape of unburnt oil into the atmosphere and so allows use of the machine in countries where stringent anti-pollution statutes are in operation. The breather tube carries the crankcase vapours to the air cleaner housing where they become mixed with the air drawn into the carburettors.

6   Excessive oil consumption indicated by blue smoke emitting from the exhaust pipes, coupled with a poor performance and fouling of spark plugs, is caused by either an excessive oil build-up in oil breather chamber, or by oil getting past the piston rings. First check the oil breather chamber and air cleaner for oil build-up. If this is the fault, check the passageway from the air/oil separator in the oil breather chamber to the lower half of

the crankcase. Blockage here will prevent oil flowing back into the crankcase, resulting in oil build-up in the breather chamber and air cleaner tube.

7   Be sure to check the oil level in the sump before starting the engine. If the oil level is not seen between the two marks adjacent to the sight window at the bottom of the clutch cover, replenish with the correct amount of oil of the specified viscosity.

10.3 The corrugated paper oil filter requires replacement at regular intervals

11 Oil pump: removal and examination

1   The oil pump is secured to the wall of the primary drive chamber behind the clutch unit. To gain access to the pump, the engine oil should be drained and the primary drive cover detached. The clutch should then be removed as described in Chapter 1, Section 12 paragraphs 3 and 4.

2   Unscrew the three screws retaining the oil pump and lift it from position. Displace the two O-rings in the casing wall. The oil pump pinion is retained on the pump shaft. Remove the circlip, lift the pinion off the shaft and push out the drive pin.

3   Remove the single screw from the reverse side of the pump body. The two halves of the pump body are located by two tight fitting dowel pins. Rather than levering the cases apart, which would damage the mating surfaces, the dowels should be driven out. Use a parallel shanked punch of a suitable size, while resting the pump across two strips of wood of a thickness sufficient to raise the pump off the workbench surface.

4   Separate the outer casing (reverse side) from the pump, leaving the drive shaft and rotors in place at this stage. Push out the drive shaft, together with the drive pin and then lift out the two rotors.

5   Wash all the pump components with petrol and allow them to dry before carrying out a full examination. Before partly reassembling the pump for the various measurements to be made, check the castings for cracks or other damage, especially the pump end covers.

6   Reassemble the pump rotors and measure the clearance between the outer rotor and the pump body, using a feeler gauge. If the clearance exceeds 0.25 mm (0.009 in) the rotor or the body must be renewed, whichever is worn. Measure the clearance between the outer rotor and the inner rotor with a feeler gauge. If this clearance is greater than 0.2 mm (0.008 in) the rotors must be renewed as a set.

7   Using a small sheet of plate glass or a straight edge placed across the pump housing, check the rotor endfloat. If the endfloat exceeds 0.15 mm (0.006 in) the complete pump must be renewed.

8   Examine the rotors and the pump body for signs of scoring, chipping or other surface damage which will occur if metallic particles find their way into the oil pump assembly. Renewal of the affected parts is the only remedy under these circumstances, bearing in mind that rotors must always be replaced as a matched set.

9   Reassemble the pump by reversing the dismantling procedure. Make sure all parts of the pump are well lubricated before the end cover is replaced and that there is plenty of oil between the inner and outer rotors. Apply a small quantity of locking fluid to the thread of the single casing screw. **Do not** omit the two O-rings when fitting the oil pump into the casing. Rotate the drive shaft as the screws are tightened down, to check that the oil pump revolves freely. A binding pump may be caused by dirt on the rotor faces or distortion of the cases, due to unequally tightened screws.

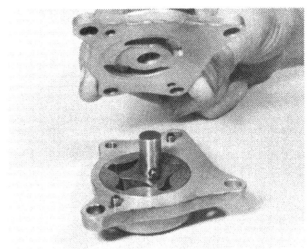

11.3 Separate the two halves of the oil pump

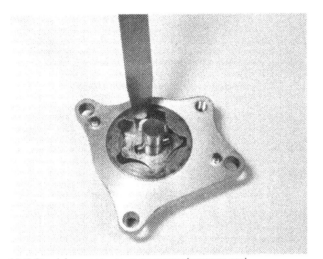

11.6 Check inner rotor to outer rotor clearance and ...

11.7 ... the rotor endfloat with a straight edge

12 Checking the oil pressure

1   Because of the predominant use of caged ball and roller bearings in the GS1000 engine a low pressure lubrication system is employed. If the condition of the oil pump is suspect, the output pressure may be checked by connecting a suitable pressure gauge to the engine.

2   A blanking plug is fitted to the right-hand end of the main oil passage which runs across the crankcase below and to the rear of the cylinder block. The blanking plug should be substituted by a suitable adaptor piece to which the pressure gauge can be attached, via a flexible hose.

3   After connection of the pressure gauge, check that the oil level in the crankcase is correct and then start the engine. The engine should be run until the oil is at approximately 60°C (140°F). Raise the engine speed to 3,000 rpm, when the pressure gauge should give a reading of 0.1 kg/cm² (1.42 psi). It can be seen that the pressure gauge must be of high sensitivity and of the correct calibration to give a useful reading. A pressure reading lower than specified may be caused by a worn oil pump or a blocked oil strainer or oil filter element. Before dismantling the pump for inspection clean the oil strainer and renew the oil filter, as described in Section 14 of this Chapter.

13 Oil pressure warning switch

1   An oil pressure failure warning switch is screwed in to a holder bolted to the top of the crankcase. The switch is interconnected with a warning light in the instrument console.

2   If the oil warning lamp comes on whilst the machine is being ridden, the engine must be stopped immediately, otherwise there is risk of severe engine failure due to a breakdown of the lubrication system. The fault must be located and rectified before the engine is re-started and run even for a brief moment.

3   Oil pressure failure may be due to a blocked oil strainer screen or a blocked filter and by-pass valve. A worn oil pump or sheared drive shaft or pin will also produce the same symptoms.

4   Failure of the switch itself is also possible. This fault should be eliminated first before seeking the cause of pressure failure elsewhere. To check the switch, it should be removed from the crankcase by unscrewing the two bolts securing it and the holding cover to the crankcase. Switch on the ignition and check that the warning bulb has illuminated. Push the brass switch plunger upwards slowly. The lamp should go out. If the warning bulb remains lit the switch is faulty and should be renewed.

4　When the engine is operated at high temperatures, there may be a tendency for the oil warning lamp to come on occasionally, at idling speeds. This is quite in order if the light extinguishes immediately engine speed is increased.

14 Oil filter: renewing the element

1　The oil filter element is contained within a semi-isolated chamber in the front of the lower crankcase, closed by a finned cover retained by three domed nuts. Before removing the cover, place a receptacle below the engine to catch the engine oil contained within the filter chamber. Drain the oil either by removing the drain plug provided or as the cover is released.

　A coil spring is fitted between the cover and the filter

element to keep the element seated firmly in position. Be prepared for the cover to fly off after removal of the bolts.
2　No attempt should be made to clean the oil filter element; it must be renewed. When renewing the filter element it is wise to renew the filter cover O-ring at the same time. This will obviate the possibility of any oil leaks.
3　The by-pass valve which allows a continued flow of lubrication if the element becomes clogged is an integral part of the filter. For this reason routine cleaning of the valve is not required since it is renewed regularly.
4　Never run the engine without the filter element or increase the period between the recommended oil changes or oil filter changes. The oil should be changed every 1500 miles and the oil filter renewed at every second oil change.

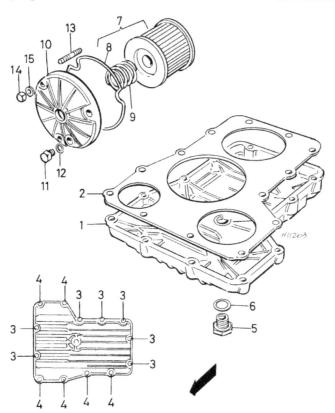

Fig. 2.6 Oil filter and sump

1　Oil sump
2　Gasket
3　Bolt – 7 off
4　Bolt – 6 off
5　Oil sump drain plug
6　Sealing washer
7　Engine oil filter assembly
8　O-ring
9　Oil filter spring
10　Oil filter cover
11　Oil filter drain plug
12　Sealing washer
13　Stud – 3 off
14　Nut – 3 off
15　Washer – 3 off

15 Fault diagnosis: fuel system and lubrication

Symptom	Cause	Remedy
Engine gradually fades and stops	Fuel starvation	Check vent hole in filler cap. Sediment in filter bowl or float chamber. Dismantle and clean
Engine runs badly. Black smoke from exhausts	Carburettor flooding	Dismantle and clean carburettor. Check for punctured float or sticking float needle.
Engine lacks response and overheats	Weak mixture Air cleaner disconnected or hose split Modified silencer has upset carburation	Check for partial block in carburettors. Reconnect or renew hose. Replace with original design.
Oil pressure warning light comes on	Lubrication system failure	Stop engine immediately. Trace and rectify fault before re-starting
Engine gets noisy	Failure to change engine oil when recommended	Drain off old oil and refill with new oil of correct grade. Renew oil filter element.

Chapter 3 Ignition system

Contents

Specifications

Alternator
Make	Nippon Denso
Type	Permanent magnet rotor, 18-coil stator
Output	250 watts @ 5000 rpm
No load voltage	More than 16V @ 5000 rpm

Ignition timing
Dwell angle	180°
Ignition timing:	
Retarded	17° BTDC below 1500 rpm
Advanced	37° BTDC above 2350 – 2500 rpm
Contact breaker gap	0.3 – 0.4 mm (0.012 – 0.016 in)

Ignition coils
Primary resistance (at 20°C)	2 – 6 ohms
Secondary resistance (at 20°C)	11 – 17 Kohms

Condenser
Capacity	0.18 microfarad

Sparking plugs
Make	NGK or Nippon Denso
Type	B-8ES or W24ES
Gap	0.6 – 0.8 mm (0.023 – 0.031 in)

1 General description

1   The spark necessary to ignite the petrol vapour in the combustion chambers is supplied by a battery and two ignition coils (one coil to two cylinders).

There are two sets of contact breaker points, two condensers, four sparking plugs and an automatic ignition advance mechanism. The contact breaker cam, which is incorporated in the advance mechanism, opens each set of points once in 180° of crankshaft rotation, causing a spark to occur in two of the cylinders. The other set of points fires 180° later, so that in every 360° of crankshaft rotation each plug is fired once. One extra spark occurs during the time when there is no combustible material in the chamber.

East set of points has one fixed and one movable contact, the latter of which pivots as the lobe of the cam separates them. The two condensers are wired in parallel, one with each set of contact points, and these function as electrical storage reservoirs, whilst also preventing arcing across the points. The condenser serves to absorb surplus current that tries to run back through the system when there is an overload situation, and feeds the current back to the ignition coils. They also help intensify the spark. When the points are closed, the current flows straight through them to earth. When they are open, there is now an open circuit. If not for the condensers, the current would arc across the points causing them to burn and pit. When the condensers reach their capacity, they discharge the current back through the primary windings and eventually to the spark plug. Any time the points get badly burnt, it is advisable to renew them, and the condensers also.

Each of the two coils has two high voltage sparking plug leads, and as in the case of points, one coil serves cylinders 1 and 4, and the other, cylinders 2 and 3.

The coils convert the low tension voltage into a high tension voltage sufficient to provide a sparking strong enough to jump the spark plug air gap. If at any time a very weak or erratic spark occurs at the plug, and the rest of the ignition system is known to be in good condition, it is time to renew an ignition coil. Although coils normally have a long life they can sometimes be faulty, especially if the outer case has been damaged.

2   The automatic advance mechanism serves to advance the ignition timing as the engine rpm rises. The mechanism is made up of two spring loaded weights which, under the action of centrifugal force created by the rotation of the crankshaft, fly apart and cause the contact points to open earlier. If the mechanism does not operate smoothly, the timing will not advance smoothly, or it may stick in one position. This will result in poor running in any but that one position. Sometimes the springs are prone to stretching, which can cause the timing to advance too soon. It is best to check the automatic advance mechanism, by carrying out a static timing test on the ignition followed by a strobe test. It is always best to check the motion of the weights by hand every 2000 miles and to clean and lubricate the unit at the same time.

3   The electrical system is powered by an ac generator (alternator) fitted to the extreme left-hand end of the crankshaft. The alternating current (ac) is passed through a full-wave rectifier where it is converted to direct currect (dc) and used to charge the battery and provide current for the lights and ancillary components. Output of the alternator is controlled by a silicon-controlled regulator (SCR unit) to within a range of 14 – 15.5 volts.

2   Crankshaft alternator: checking the output

1   If the charging performance of the alternator is suspect, it can be checked with a multi-meter test instrument that includes a voltmeter and ohmmeter. As most owner/riders are unlikely to possess equipment of this type it is advised that the machine be returned to a Suzuki Service Agent for testing.

2   If a multi-meter is available, an initial check on the alternator and the rectifier and regulator assemblies may be carried out as described in Chapter 6. As mentioned in Chapter 6, Section 3 the charging system should be considered as a whole, and should be tested accordingly.

3   Ignition coils: checking

1   Each ignition coil is a sealed unit, designed to give long service without need for attention. They are located within the top frame tubes, immediately to the rear of the steering head assembly. If a weak spark and difficult starting causes the performance of a coil to be suspect, it should be tested by a Suzuki Service Agent or an auto-electrical engineer who will

have the appropriate test equipment. A faulty coil must be renewed; it is not possible to effect a satisfactory repair.

2   A defective condenser in the contact breaker circuit can give the illusion of a defective coil and for this reason it is advisable to investigate the condition of the condenser before condemning the ignition coil. Refer to Section 6 of this Chapter for the appropriate details.

3   Note that it is extremely unlikely that both ignition coils will prove faulty at the same time, unless the common electrical feed is in some way deranged. This can be checked by measuring the low tension voltage supplied to the coils, using a voltmeter.

4   Contact breaker: adjustments

1   To gain access to the contact breaker assembly, it is necessary to detach the aluminium cover retained by three crosshead screws at the righthand end of the crankshaft. Note that the cover has a sealing gasket, to prevent the ingress of water.

2   Rotate the engine slowly by means of the engine turning hexagon until one set of points is in the fully open position. Examine the faces of the contacts. If they are blackened and burnt, or badly pitted, it will be necessary to remove them for further attention. See Section 5 of this Chapter. Repeat for the second set of contact points.

3   Adjustment is effected by slackening the screw through the plate of the fixed contact breaker point and moving the point either closer to or further from the moving contact until the gap is correct as measured by a feeler gauge. The correct gap with the points FULLY OPEN is 0.3 – 0.4 mm (0.012 – 0.016 in). Small projections on the contact breaker baseplate permit the insertion of a screwdriver to lever the adjustable point into its correct location. Repeat this operation for the second set of points, which must also be fully open.

4   Do NOT slacken the two screws through the extremities of the larger baseplate fitted to the right hand set of contact breaker points. They are used for adjusting the setting of the ignition timing and it will be necessary to re-time the engine if the baseplate is permitted to move. Only the centre screw should be slackened, to adjust the fixed contact breaker point.

5   Before replacing the cover and gasket, place a light smear of grease on the contact breaker cam and one or two drops of thin oil on the felt which lubricates the surface of the cam. It is better to under-lubricate rather than add excess because there is always chance of excess oil reaching the contact breaker points and causing the ignition circuit to malfunction.

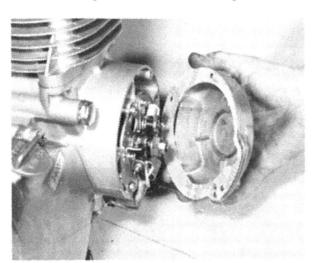

4.1 Detach the points cover and check condition of the sealing gasket

4.3a Slacken single screw to adjust each set of points and ...

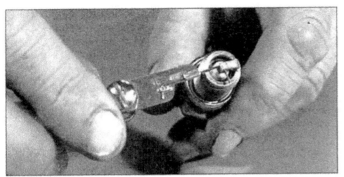

Electrode gap check - use a wire type gauge for best results

Electrode gap adjustment - bend the side electrode using the correct tool

Normal condition - A brown, tan or grey firing end indicates that the engine is in good condition and that the plug type is correct

Ash deposits - Light brown deposits encrusted on the electrodes and insulator, leading to misfire and hesitation. Caused by excessive amounts of oil in the combustion chamber or poor quality fuel/oil

Carbon fouling - Dry, black sooty deposits leading to misfire and weak spark. Caused by an over-rich fuel/air mixture, faulty choke operation or blocked air filter

Oil fouling - Wet oily deposits leading to misfire and weak spark. Caused by oil leakage past piston rings or valve guides (4-stroke engine), or excess lubricant (2-stroke engine)

Overheating - A blistered white insulator and glazed electrodes. Caused by ignition system fault, incorrect fuel, or cooling system fault

Worn plug - Worn electrodes will cause poor starting in damp or cold weather and will also waste fuel

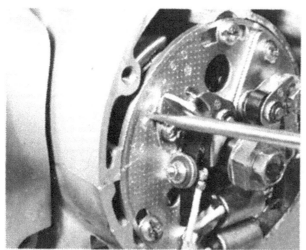

4.3b ... lever the adjustable point by using the projections on the baseplate

4.4 Do NOT slacken the two outermost screws (arrowed), see text, unless timing is to be adjusted

5 Contact breaker points: removal, renovation and replacement

1   If the contact breaker points are burned, pitted or badly worn, they should be removed for dressing. If it is necessary to remove a substantial amount of material before the faces can be restored, the points should be renewed.

2   To remove the contact breaker points, detach the circlip which secures the moving contact to the pin on which it pivots. Remove the nut and bolt which secures the flexible lead wire to the end of contact return spring, noting the arrangement of the insulating washers so that they are replaced in their correct order during reassembly. Lift the moving contact off the pivot, away from the assembly.

3   The fixed contact is removed by unscrewing the screw which retains the contact to the contact breaker baseplate.

4   The points should be dressed with an oilstone or fine emery cloth. Keep them absolutely square throughout the dressing operation, otherwise they will make angular contact on reassembly, and rapidly burn away.

5   Replace the contacts by reversing the dismantling procedure, making sure that the insulating washers are fitted in the correct order. It is advantageous to apply a thin smear of grease to the pivot pin, prior to replacement of the moving contact arm.

6   Check, and if necessary, re-adjust the contact breaker gap when the points are fully open. Repeat the whole operation for the second set of points.

6 Condensers: removal and replacement

1   A condenser is included in each contact breaker circuit to prevent arcing across the contact breaker points as they separate. It is connected in parallel with each set of points and if a fault develops, ignition failure is liable to occur.

2   If the engine proves difficult to start, or misfiring occurs, it is possible that the condenser is at fault. To check, separate the contact breaker points by hand when the ignition is switched on. If a spark occurs across the points and they have a blackened and burnt appearance, the condenser can be regarded as unserviceable.

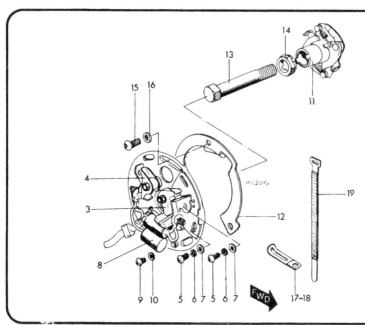

Fig. 3.1 Contact breaker assembly

1   Contact breaker assembly
2   Contact breaker plate
3   Right-hand contact breaker unit
4   Left-hand contact breaker unit
5   Screw – 4 off
6   Spring washer – 4 off
7   Washer – 4 off
8   Condenser – 2 off
9   Screw – 2 off
10 Spring washer – 2 off
11 Automatic timing unit (ATU)
12 Ignition timing index plate
13 Bolt
14 Engine turning hexagon
15 Screw – 3 off
16 Washer – 3 off
17 Cable clip
18 Cable clip
19 Cable strap

3 It is not possible to check a condenser without the appropriate test equipment. In view of the low cost involved, it is preferable to fit a new one and observe the effect on engine performance.

4 Because each condenser and its associated set of contact breaker points is common to a pair of cylinders, a faulty condenser will not cause a misfire on one cylinder only. In such a case it is necessary to seek the cause of the trouble elsewhere, possibly in some other part of the ignition-circuit or the carburettor. It also follows that both condensers are unlikely to fail at the same time unless damaged in an accident. If the cases are crushed or dented, electrical breakdown will occur.

5 The condensers are located at the base of the contact breaker assembly, parallel to each other. Each has an integral bracket and is attached to the contact breaker baseplate by a single crosshead screw, making renewal easy.

7 Ignition timing: checking and resetting

1 In order to check the accuracy of the engine timing, it is necessary to remove the contact breaker cover from the right-hand side of the crankcase. Ignition timing checking and resetting should take place after resetting the contact breaker gaps, as described in Section 4.

2 Apply a spanner to the engine rotation hexagon and turn the engine in a forward direction, whilst viewing the ATU through the inspection aperture in the contact breaker stator plate. It will be seen that there is a set of three scribed lines on each side of the ATU.

3 Commence ignition timing checking on the left-hand contact breaker set, which controls cylinders No 1 and 4. To determine at which moment the points open connect a 12v bulb between the moving point and a suitable earthing point on the engine. With the ignition turned on, the bulb will light up when the points are open. Rotate the engine until the F1-4 mark on the ATU is in **exact** alignment with the index pointer mark on the plate fitted to the rear of the stator plate. If the ignition is correct, the points should be on the verge of opening when this position is reached. This will be indicated by the flickering of the bulb.

4 To adjust the ignition timing on No 1 and 4 cylinders slacken the three screws which pass through the elongated holes in the stator plate periphery. Rotate the plate until the light flickers and then tighten the screws. Turn the engine backwards about 90° and then forwards again to check the setting.

5 Check the ignition timing on Nos 2 and 3 cylinders in a similar manner, using the F2-3 timing mark. If the timing is incorrect, slacken the two screws holding the right-hand

contact breaker assembly mounting plate to the main stator plate. Move the plate to the correct position and tighten the screws. Recheck the timing.

6 Provided that the contact breakers are in good condition and care is taken, manual adjustment of the ignition timing should be acceptably accurate. If possible, however, the timing should be checked using a stroboscope lamp because not only can the accuracy of the timing be checked with the engine running but the correct performance of the ATU can be verified. The timing light should be connected to the low tension or high tension side of the ignition as instructed by the manufacturers of the light. Test the left-hand contact breaker and then the right-hand contact breaker. Start the engine and illuminate the ATU through the inspection aperture. With the engine running below 1500 rpm the F mark should be in alignment with the index mark. Raise the engine speed slowly to 2500 rpm when the advance mark should align with the index pointer.

7 The advance range of 1000 rpm peaks at 2500 rpm, above which engine speed no more advance is possible. If, when increasing the engine speed from the commencement of advance at 1500 rpm, the timing marks are seen to move erratically, or if the advance range has altered appreciably, the ATU should be inspected for wear or malfunctioning as described in the following Section.

6.5 Condensers are held by a single screw each below the points assembly

7.5a Timing adjustment screws for cylinders 2 and 3 (arrowed), and index marks on back plate align with those on ATU

7.5b Use slot provided to alter plate position

8 Automatic timing unit: examination

1   The automatic timing unit rarely requires attention although it is advisable to examine it periodically.

2   To obtain access to the unit remove the inspection cover and the contact breaker back plate complete with contact breakers. The ATU centre bolt and engine turning hexagon should be removed before the stator plate. Before removal, the back plate and end cover should be marked so that the back plate can be replaced in exactly the same position. This will ensure the ignition timing is not altered.

3   Pull the ATU from position, noting the drive pin with which it locates and is driven. The unit comprises balance weights which move outwards against spring tension as the centrifugal forces increase. The balance weights must move freely on their points, which should be lubricated. The tension springs must also be in good condition.

4   Check the surface of the contact breaker cam for pitting or obvious signs of wear. Damage to the cam cannot be rectified; the complete ATU must be renewed.

5   When replacing the ATU, check that the drive pin engages with the recess in the rear of the centre boss. Because there is a single recess only, the ATU cannot inadvertently be replaced in the incorrect position and so alter the timing marks in relation to the crankshaft.

8.3 Check bob-weights, pivots and springs on ATU

9 Sparking plugs: checking and resetting the gaps

1   All models are fitted with Nippon Denso type W24ES or NGK type B-8ES sparking plugs as standard, gapped within the range 0.6 – 0.8 mm (0.024 – 0.031 in). Operating conditions may indicate a change in sparking plug grade; the type recommended by the manufacturer gives the best, all round service.

2   Check the gap of the plug points during every four monthly or 4000 mile service. To reset the gap, bend the outer electrode to bring it closer to the centre electrode and check that a 0.6 mm (0.024 in) feeler gauge can be inserted. Never bend the central electrode or the insulator will crack, causing engine damage if the particles fall in whilst the engine is running.

3   With some experience, the condition of the sparking plug electrodes and insulator can be used as a reliable guide to engine operating conditions.

4   Beware of overtightening the sparking plugs, otherwise there is risk of stripping the threads from the aluminium alloy cylinder heads. The plugs should be sufficiently tight to sit firmly on their copper sealing washers, and no more. Use a spanner which is a good fit to prevent the spanner from slipping and breaking the insulator.

5   If the threads in the cylinder head strip as a result of over tightening the sparking plugs, it is possible to reclaim the head by the use of a Helicoil thread insert. This is a cheap and convenient method of replacing the threads; most motorcycle dealers operate a service of this kind.

6   Make sure the plug insulating caps are a good fit and have their rubber seals. They should also be kept clean to prevent tracking. These caps contain the suppressors that eliminate both radio and TV interference.

10 Fault diagnosis: ignition system

Symptom	Cause	Remedy
Engine will not start	Faulty ignition switch	Operate switch several times in case contacts are dirty. If lights and other electrics function, switch may need renewal.
	Starter motor not working	Discharged battery. Remove and recharge battery.
	Short circuit in wiring	Check whether fuse is intact. Eliminate fault before switching on again.
	Completely discharged battery	If lights do not work, remove battery and recharge.
Engine misfires	Faulty condenser in ignition circuit	Renew condenser and re-test.
	Fouled sparking plug	Renew plug and have original cleaned.
	Poor spark due to generator failure and discharged battery	Check output from generator. Remove and recharge battery.
Engine lacks power and overheats	Retarded ignition timing	Check timing and also contact breaker gap. Check whether auto-advance mechanism has jammed.
Engine 'fades' when under load	Pre-ignition	Check grade of plugs fitted; use recommended grades only.

Chapter 4 Frame and forks

Contents

Specifications

Frame
Type .... Tubular, double cradle

Front forks
Type .... Telescopic, hydraulically damped with variable air assistance
Damping oil capacity (All USA models except S) .... 240 cc (8.15/6.76 US/Imp fl oz) per leg
(All UK models and S [USA]) .... 260 cc (8.75/7.32 US/Imp fl oz) per leg
Damping oil grade .... SAE 10W/20 or fork oil
Damping oil level .... 140 mm (5.51 in) from top of stanchion
Fork spring free length .... 421 mm (16.57 in)
Service limit .... 416 mm (16.38 in)
Fork air pressure:
   Standard setting .... 0.8 kg cm² (11 psi)
   Maximum pressure .... 2.5 kg cm² (35 psi)*

Rear suspension
Type .... Swinging arm, welded tubular steel

Rear suspension units
Type .... Hydraulically damped, 5-way adjustable spring preload, and
   4-way adjustable damping
EC and EN (UK spec) models .... Hydraulically damped, with variable air assistance
Suspension unit air pressure (EC and EN):
   Standard setting .... 2.0 – 2.2 kg cm² (28.5 – 31.0 psi)
   Maximum pressure .... 3.4 kg/cm² (48.5 psi)*

*Not suitable for a prolonged length of time

Main torque wrench settings
Steering head side pinch bolt (8 mm) .... 1.5 – 2.5 kgf m (10.8 – 18 lbf ft)
Steering head centre bolt (12 mm) .... 3.6 – 5.2 kgf m (26 – 37 lbf ft)
Upper yoke pinch bolts (10 mm) .... 2.0 – 3.0 kgf m (14 – 21 lbf ft)
Lower yoke pinch bolts (8 mm) .... 1.5 – 2.5 kgf m (11 – 18 lbf ft)
Swinging arm pivot shaft nut (16 mm) .... 5.0 – 8.0 kgf m (36 – 57 lbf ft)
Rear suspension units:
   Upper mounting nut (10 mm) .... 2.0 – 3.0 kgf m (14 – 21 lbf ft)
   Lower mounting bolt (10 mm) .... 2.0 – 3.0 kgf m (14 – 21 lbf ft)

1 General description

The Suzuki GS1000 models all share the same conventional duplex cradle frame. This type of frame has the engine suspended inside the frame tubes rather than the engine comprising any part of the frame.

The suspension on these models does not, however, follow convention to the same extent as the frame. The front forks, whilst being of the usual telescopic type, have unusual internal damping arrangements. The damping medium is still predominantly oil, but with the added benefit of variable air assistance. Conventional fork springs are also contained within each fork stanchion. The use of air, enables the damping characteristics of the front forks to be adjusted, almost infinitely, to suit particular riders and riding conditions, by increasing or reducing the pressure

In addition to the above, the GS1000 L (Low Slinger) model is fitted with front forks that carry the front spindle forward of the stanchions. This fork design, usually referred to as leading-axle, provides extra travel and conforms to the overall styling image presented in the L model.

Air is added to the front forks by means of a Shraeder valve, enclosed beneath a chrome protector cap, at the top of each fork leg. Each fork leg can be detached from the machine as a complete unit, without dismantling the steering head assembly.

Rear suspension is provided by a swinging arm fork, pivoting on a shaft and two needle roller bearings. Rear springing and damping is effected by the use of two types of adjustable suspension units. The E models, as supplied to the UK, are equipped with sealed, air adjustable rear units and as such offer numerous damping options. The rear suspension units on all other models incorporate 5-way adjustment of the spring preload and a further 4-way rebound damping adjustment facility. The latter adjustment is facilitated by the provision of a knurled ring setting wheel at the top of each rear unit. A total of twenty different combinations of spring preload and damping force are available to cope will all riding requirements.

2   Front forks: removal from the frame

1   It is unlikely that the front forks will need to be removed from the frame as a complete unit, unless the steering head bearings require attention or the forks are damaged in an accident. In the event of damage to one or both of the fork legs, they may be removed from the steering head and lower fork yoke, whilst leaving the two latter components in situ on the frame.

2   On GS1000 S models, the first stage in a complete fork removal operation is the detaching of the fairing. Remove the four bolts that retain the lower part of the fairing to the bottom fork yoke. Slide the fairing away forwards, to clear the headlamp, and ensuring no wires or cables are caught up, remove the complete fairing.

3   On all models the control cables from the handlebar levers can be removed. Alternatively, remove the levers complete with the cables attached. The shape of the handlebars fitted and the length of control cables will probably dictate the method used. Remove the front brake lever master cylinder unit, which is retained by a clamp held by two bolts. Tie the master cylinder to some part of the machine not to be dismantled, so that it is secure and resting in an upright position.

4   Detach the handlebars from their mounting points on the fork upper yoke. The handlebars are held by two U-clamps, retained by two bolts and spring washers each. On GS1000 E (EC and EN) and S models, there is a separate plastic housing fitted over the four handlebar mounting bolts and the two clamps they retain. By removing the plastic housing, access to the same mounting clamp and bolt arrangement is obtained. Remove the headlamp unit from the headlamp shell and disconnect the electrical leads at the snap connectors. No difficulty should be encountered in replacement, as the connections are mainly of the block type and the wiring is colour coded.

5   Disconnect the speedometer and tachometer cables at the instrument, where they are retained by knurled rings. Detach the wiring connections from the instrument bulb holders and from the warning lamp console at the two separate block connectors. Undo the two bolts which pass through the instrument mounting bracket and lift the complete unit from the machine. Detach all further connections, for the extra instrumentation which is fitted to the GS1000 S model. On all models the fuel gauge leads must be disconnected

6   Remove the front wheel as described in Chapter 5, Section 3. Where two caliper units are fitted, both should be detached from the fork legs. Temporarily, tie each to the frame down tubes so that their weight is not taken by the hydraulic hoses.

7   Remove the front mudguard, which is retained by two bolts passing into each fork leg.

8   Remove the chrome valve protector caps at the top of each fork leg. Release all the air pressure from the forks by depressing the centre of the Shraeder valve.

9   Loosen the clamp bolts which retain the fork legs in the upper and lower yokes. The fork legs can now be eased downwards, out of position. If the clamps prove to be excessively tight, they may be gently sprung, using a large screwdriver. This must be done with great care, to prevent breakage of the clamps, necessitating renewal of the complete yoke. Alternatively, if the fork legs appear to be binding in their clamps, such as may occur if they are bent or twisted generally, as the result of an accident, liberal use of washing-up liquid applied to the rubber seal in the clamps, can be very beneficial. When removing the second leg, care should be taken to support the headlamp, and then, once freed, lift it away from the machine.

10   Remove the single bolt which secures the hydraulic hose union to the fork lower yoke. The complete front brake assembly, including the master cylinder, hoses and calipers, may be lifted away from the machine without the need to drain the fluid. By following this procedure, draining, refilling, and bleeding of the system is not required.

11   Loosen the clamp bolt located at the rear of the upper yoke, and from the top of the yoke remove the large chrome bolt together with the washer. From the underside, tap the upper yoke upwards until it frees the steering column. Support the weight of the lower yoke and, using a C-spanner, remove the steering head bearing adjuster ring. If a C-spanner is not available, a soft brass drift and hammer may be used to slacken the nut.

12   Remove the upper dust seal and outer race (cone) once the adjuster nut has been detached. The bottom yoke, complete with steering column, can now be lowered from position. Unlike the majority of present-day motor cycles, the GS1000 range is fitted with taper roller bearings, suitable caged, in the steering head. The lower bearing race will be lowered from position as the steering column is removed. Separate the bearing from the column, noting the positioning of a heavy washer below the lower race.

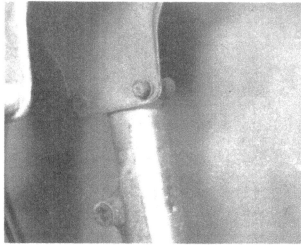

2.7 The mudguard is held by two bolts on each leg

2.8 Detach the chromed valve cover and release pressure

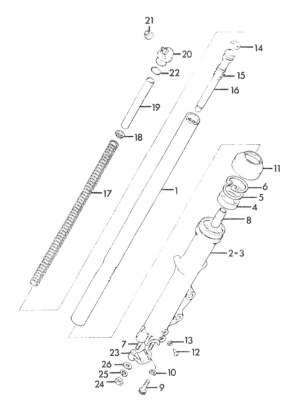

Fig. 4.1 Front forks

1	Inner tube (stanchion) 2 off	14	Piston ring – 2 off
2	RH lower leg	15	Rebound spring – 2 off
3	LH lower leg	16	Damper rod – 2 off
4	Oil seal – 2 off	17	Fork spring – 2 off
5	Washer – 2 off	18	Spring seat – 2 off
6	Circlip – 2 off	19	Collar – 2 off
7	Stud – 4 off	20	Cap – 2 off
8	Damper rod seat – 2 off	21	Valve cap – 2 off
9	Allen screw – 2 off	22	O-ring – 2 off
10	Sealing washer – 2 off	23	Clamp – 2 off
11	Dust excluder – 2 off	24	Nut – 4 off
12	Drain screw – 2 off	25	Lock washer – 4 off
13	Sealing washer – 2 off	26	Plain washer – 4 off

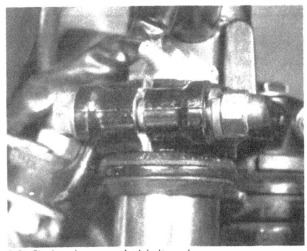

2.9a Slacken the upper pinch bolts and ...

2.9b ... then the lower yoke pinch bolts

3   Front forks: dismantling

1   It is advisable to dismantle each fork leg separately, using an identical procedure. There is less chance of unwittingly exchanging parts if this approach is adopted. Commence by draining each fork leg of damping oil; there is a drain plug in each lower leg above and to the rear of the wheel spindle housing. Remove the valve and then the valve cap unit, noting the O-ring below the valve cap unit. The fork has one long spring, a spacer, and a spring guide. This can now be removed.

2   Clamp the fork lower leg in a vice fitted with soft jaws, or wrap a length of rubber inner tube around the leg to prevent damage. Unscrew the socket screw, recessed into the housing which carries the front wheel spindle. Prise the dust excluder from position and slide it up the fork upper tube. The upper tube (stanchion) can be pulled out of the lower fork leg. Pull the damper rod seat off the rod, and separate the damper rod from the stanchion. A simple device consisting of a bolt of the correct head size and two nuts inserted in the damper rod end and on the extension bar and socket, enable the damper rod to be pulled out of position towards the top end of the tube.

3 The oil seal fitted to the top of the lower leg should be removed only if it is to be renewed. Renewal of the oil seals is the only course of action if their ability to contain the fork oil has become impaired. With the air-assisted forks fitted to the GS1000, a secondary function of oil seals has to be considered. They must be capable of containing the required air pressure, without any pressure losses. This is particularly important in view of the small amount of pressure involved in the normal operation of the forks. It follows that with only a small quantity of air pressurized, it only needs a slight leakage to seriously affect the amount contained within the forks. It is unlikely that the seals would fail simultaneously and therefore an imbalance would occur in damping which might seriously affect handling. The reason for not disturbing the oil seal unless failure has occured is that damage will almost certainly be inflicted when it is prised from position. The seal is retained by a washer and a circlip.

4 Note the copper sealing washer fitted to the socket head bolt at the fork leg bottom. Ensure it is reinstalled on reassembly.

3.1a Remove the valve and ...

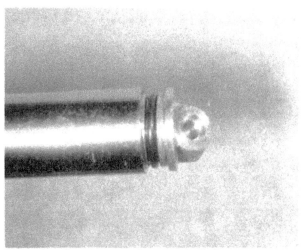

3.1b ... then the valve cap unit, noting the O-ring

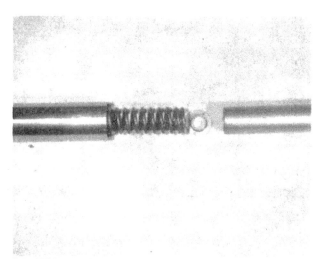

3.1c Withdraw the fork spring, spacer and spring guide

3.2a Unscrew the socket bolt from the lower leg ...

3.2b ... and remove it, noting the sealing washer

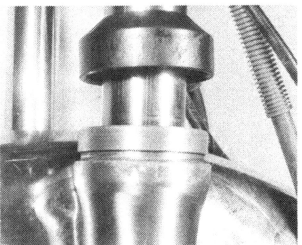

3.2c Prise off the dust cover

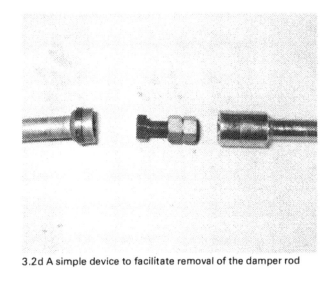

3.2d A simple device to facilitate removal of the damper rod

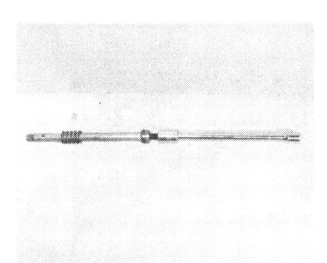

3.2e The device being used in conjunction with an extension bar and socket

3.3a Removal of the circlip and washer enables ...

3.3b ... the oil seal to be prised from position if renewal is required

4 Front forks: examination and renovation

1 The front forks do not contain bushes. The fork legs slide directly against the hard chrome surface of the fork tubes. If wear occurs, indicated by slackness, the fork leg complete will have to be renewed, possibly also the fork stanchion. Wear on the fork stanchion is indicated by scuffing and penetration of the hard chrome surface.

2 After an extended period of service the fork spring may take a permanent set. The service limit for the fork springs is 416 mm. Always fit new springs as a pair, NEVER separately.

3 Check the outer surface of the stanchion for scratches or roughness. It is only too easy to damage the oil seal during reassembly, if these high spots are not eased down. The stanchions are unlikely to bend unless the machine is damaged in an accident. Any significant bend will be detected by eye, but if there is any doubt about straightness, roll the stanchion tubes on a flat surface. If the stanchions are bent, they must be renewed. Unless specialised repair equipment is available, it is rarely practicable to effect a satisfactory repair to a damaged stanchion.

4 The piston ring fitted to the damper rod may wear if oil changes at the specified intervals are neglected. If damping has

become weakened and does not improve as a result of an oil change, the piston ring should be renewed. Check also that the oilways in the damper rod have not become obstructed.

5 Steering head bearings: examination and renovation

1 Clean off any dirt that has accumulated on the upper and lower bearing races. They should also be thoroughly examined for any signs of wear or damage. No signs of indentation, and a polished appearance should be apparent. Renew the set if necessary.

2 The roller bearings themselves should be cleaned and examined.Check the rollers for any signs of damage, such as scoring marks, and check for a uniform polished appearance. If any require replacement the whole set must be renewed.

4 The outer races are a drive fit in the steering head lug and may be drifted out,using a suitable long handled drift passed through the centre of the lug. The lower inner race may be levered from position on the steering stem. When driving the new outer races into place, ensure that they remain square to the housing in the lug or the housing may be damaged. Apply a small amount of grease to each set of bearings before they are reinstalled in the steering head.

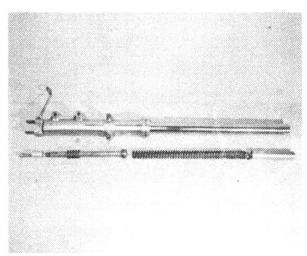

4.1 Check all components before reassembly

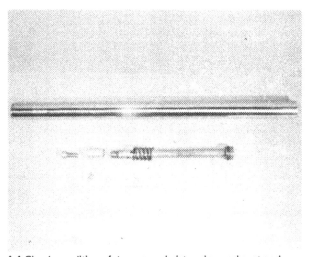

4.4 Check condition of damper rod piston ring, rod seat and oilways in the rod

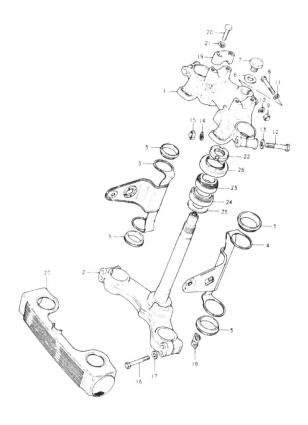

Fig. 4.2 Steering head assembly

1 Upper (crown) yoke	15 Domed nut – 2 off
2 Lower yoke/steering stem	16 Pinch bolt – 4 off
3 RH headlamp bracket	17 Washer – 4 off
4 LH headlamp bracket	18 Double nut – 2 off
5 Bracket guide – 4 off	19 Handlebar clamp – 2 off
6 Washer	20 Bolt – 4 off
7 Crown bolt	21 Spring washer – 4 off
8 Pinch bolt	22 Bearing adjuster ring
9 Domed nut	23 Upper bearing
10 Spring washer	24 Lower bearing
11 Washer	25 Washer
12 Pinch bolt – 2 off	26 Dust excluder
13 Washer – 2 off	27 Cover
14 Spring washer – 2 off	

6 Front forks: replacement

1 Replace the front forks by following in reverse the dismantling procedures described in Sections 2 and 3 of this Chapter. Before fully tightening the front wheel spindle clamps and the fork yoke pinch bolts,bounce the forks several times to ensure they work correctly and settle down into their original settings. Complete the final tightening from the wheel spindle upwards.

2 Refill each fork leg with the correct quantity and specification of fork oil before replacing the handlebars. The handlebars obstruct access to the filler orifices. Suzuki recommend that the damping fluid be of SAE 10W/20 viscosity. Check that the drain plugs have been re-inserted and tightened before the oil is added.

3 If the fork stanchions prove difficult to relocate through the

fork yokes, make sure their outer surfaces are clean and polished, and additionally, use a liberal coating of washing-up liquid, to ease their relocation. This will allow the stanchions to slide more easily. It can also often be advantageous to use a screwdriver blade to open up the clamps, as the tubes are moved upwards into position.

4 Before the machine is used on the road, check the adjustment of the steering head bearings. If they are too slack, judder will occur especially during braking. There should be no detectable play in the head races when the handlebars are pulled and pushed with the front brake applied hard.

5 Overtight head races are equally undesirable. It is possible to unwittingly apply a loading of several tons on the head races when they have been overtightened, even though the handlebars appear to turn quite freely. Overtight bearings will made the machine roll at low speeds and give generally imprecise handling with a tendency to weave. Adjustment is correct if there is no perceptible play in the bearings and the handlebars will swing to full lock in either direction, when the machine is on the centre stand with the front wheel clear of the ground. Only a slight tap should cause the handlebars to swing.

6 Similarly, before atempting to ride the machine, the forks must be pressurized with the required amount of air. With the valves installed in the top of each fork leg, but not the protector caps, proceed to pump up to the required pressure. Suzuki recommend the use of a bicycle-type pump, with a suitable adaptor to accept the Shaeder valve, and their air pressure gauge (Suzuki part No. 09940-44110). A garage air line should only be used as an emergency, as it is not possible to meter the amount of air entering the forks with enough control to gain the correct pressure reading. There is also a good chance that the fork seals will fail due to the sudden excessive pressure exerted upon them. The front wheel must be clear of the ground when carrying out this process.

7 Note that only air or nitrogen should be used to pressurize the forks. Never attempt to use oxygen or any other gas which explodes under pressure. Suzuki recommend a pressure of 0.8 kg/cm^2 (11 psi) as a base figure from which to adjust the level of air-assistance, according to the operators' personal preference. The pressure can be allowed to rise to 2.5 kg/cm^2 (35 psi) safely, for a short space of time. The normal working pressure range, however, is between 0.8 and 1.2 kg cm^2 (11 to 18 psi). Exceeding the maximum specified pressure loading will damage fork seals and possibly promote actual fork damage. Always ensure that the difference between the two fork leg pressures is minimal, and never more than 0.1 kg/cm^2 (1.4 psi).

8 When adjusting the front fork air pressure, it must be borne in mind that the rear suspension must also be adjusted to preserve the balance between front and rear damping. It is possible, through total mismatching of the operation of front and rear suspensions, to upset the normal geometry of the frame, to such an extent as to render the machine highly unstable. See Section 10 of this Chapter for rear suspension adjustment procedure.

7 Steering head lock

1 The steering head lock on GS1000 models is incorporated in the ignition switch mounted on the upper fork yoke between the speedometer and tachometer. The lock, when in a locked position, has a tongue which extends from the body of the lock, when the handlebars are turned to the left or right, and abuts against a plate welded to the base of the steering head. The lock is operated when the ignition switch is turned to the lock or park position.

2 If the lock malfunctions it must be renewed; a repair is impracticable. When the lock is changed the key must be changed too, to match the new lock.

8 Frame: examination and renovation

1 The frame is unlikely to require attention unless accident damage has occurred. In some cases, replacement of the frame is the only satisfactory course of action if it is badly out of alignment. Only a few frame repair specialists have the jigs and mandrels necessary for resetting the frame to the required standard of accuracy and even then there is no easy means of assessing to what extent the frame may have been over-stressed.

2 After the machine has covered a considerable mileage, it is advisable to examine the frame closely for signs of cracking or splitting at the welded joints. Rust can also cause weakness at these joints. Minor damage can be repaired by welding or brazing, depending on the extent and nature of the damage.

3 Remember that a frame which is out of alignment will cause handling problems and may even promote speed wobbles. If misalignment is suspected, as the result of an accident, it will be necessary to strip the machine completely so that the frame can be checked and, if necessary, renewed.

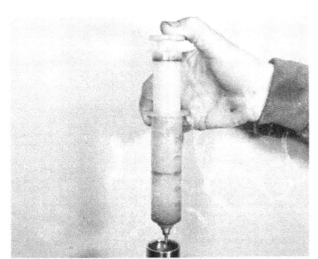

6.2 Fill each leg with the correct quantity of fluid

6.6 Use pump connected to valve at top of each leg to fill with air pressure

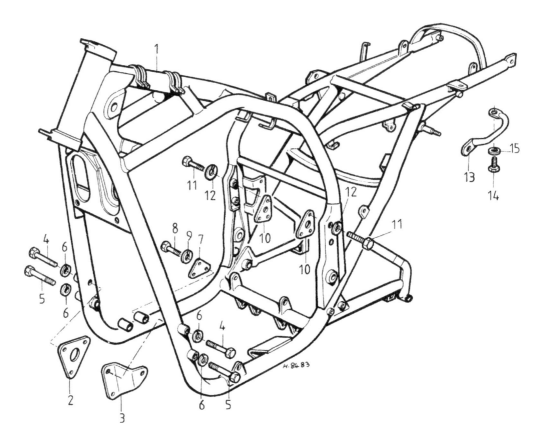

Fig. 4.3 Frame

1	Frame	6	Spring washer – 4 off	11	Bolt – 4 off
2	Right-hand engine mounting plate	7	Engine mounting plate	12	Spring washer – 4 off
3	Left-hand engine mounting plate	8	Bolt – 2 off	13	Grab handle
4	Bolt – 2 off	9	Spring washer – 2 off	14	Bolt
5	Bolt – 2 off	10	Rear engine mounting – 2 off	15	Washer

9 Swinging arm fork: dismantling and renovation

1   The rear fork of the frame is of the swinging - arm type. It pivots on a shaft that passes through the crossmember and both sides of the main frame assembly. The pivot shaft assembly itself consists of a centre spacing collar, two taper roller bearings, two smaller spacers, and a washer and dust seal at each end. The whole assembly is held together by a washer and large nut on the left-hand end of the shaft. Worn swinging arm bushes can be detected by placing the machine on its centre stand and pulling and pushing vigorously on the rear wheel in a horizontal direction. Any play will be noticeable by the leverage effect.

2   When wear develops in the swinging arm, necessitating renewal of the bearings, the renovation procedure is quite straightforward. Commence by removing the rear wheel as described in Chapter 5, Section 10.

3   Suspend the caliper unit from the right-hand rear indicator stalk using a length of wire or string. If care is taken during dismantling, disconnection of the rear brake hydraulic pipe will not be required. This will facilitate reassembly. Carefully ease the hydraulic pipe and the grommet through which it passes from position in the locating clip welded to the swinging arm.

4   Remove the lower two bolts that hold the suspension units to the swinging fork, so that the fork swings down. Leave the

suspension units hanging from the frame, but slacken the top nut so that they are free to move. This facilitates reassembly.

5   Take out the swinging arm pivot shaft by undoing the nut on the left-hand side of the machine. This may need a gentle tap with a rawhide mallet and drift to displace it. Pull the final drive chain across so that it clears the swinging arm fork left-hand end. The swinging arm is now free and can be lifted out to the rear.

6   Remove the dust cap and thrust washer from each side of the swinging arm cross-member and then pull out the two short spacers. Push out the long central spacer, using a long shanked screwdriver.

7   The caged needle roller bearings may be drifted out of position, using a suitable length of steel rod. Do not remove the bearings merely for inspection as the cages will be damaged by the drift. Lubricate the outside of the new bearing cages before driving them into place, and ensure that they are fitted with the punch marked face outwards.

8   Check that the swinging arm pivot shaft is straight. A bent shaft may be straightened in a jig. If this cannot be accomplished, the shaft should be renewed.

9   Reassemble the swinging arm fork by reversing the dismantling procedure. Grease the pivot shaft and bearings liberally prior to reassembly, bearing in mind that no provision is given for subsequent lubrication when the swinging arm is fitted to the machine.

9.4a Remove the lower rear suspension mounting bolts and ...

9.4b ... slacken the top mounting nuts

9.5a After removal of the pivot shaft and nut ...

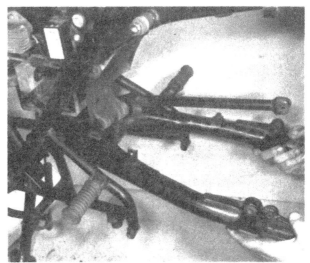

9.5b ... lift the swinging arm fork out upwards and rearwards

9.6a Remove the dust cap and shim from the cross-member

9.6b Pull out the short inner bush and ...

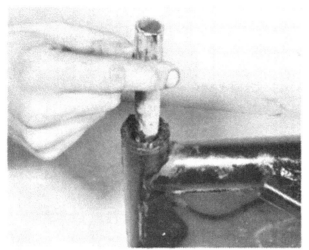

9.6c ... the longer central spacer

9.9 Grease the swinging arm bearings liberally before refitting dust cap

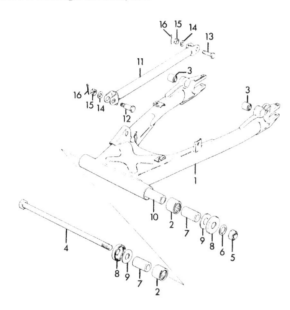

Fig. 4.4 Swinging arm assembly

1	Swinging arm	9	Shim – A/R
2	Bearing – 2 off	10	Centre spacer
3	Bonded rubber bush	11	Torque arm
4	Pivot shaft	12	Shouldered bolt
5	Nut	13	Bolt
6	Washer	14	Washer – 2 off
7	Inner bush – 2 off	15	Castellated nut – 2 off
8	Dust cap – 2 off	16	Split pin – 2 off

10 Rear suspension units: examination and adjustment

1   Rear suspension units of the hydraulically damped type are fitted to the Suzuki GS models. They can be adjusted to give five different spring loadings, without removal from the machine. In addition to this adjustment, the rear suspension on the GS1000 models has a 4-way damping adjustment facility.
2   Each rear suspension unit has a peg hole above the adjusting notches, to facilitate spring tension adjustment. Either a C-spanner or the screwdriver supplied with the original tool kit can be used to turn the adjusters. Turn clockwise to increase the spring tension and so stiffen up the suspension springing. The recommended settings are:-

Position 1 (least tension) for normal solo riding and
Position 5 (greatest tension) for high speed riding and/or when carrying a heavy load.

The intermediate positions offer additional settings for varying conditions, as required.
3   The adjustment of the damping force of the rear suspension is carried out in a similar fashion to the spring tension adjustment. A rubber cap is fitted to the top of each suspension unit. Prise this free from its normal position to reveal the slotted damper adjustment wheel. Insert the blade of a screwdriver into one of the adjuster slots and turn clockwise to increase the amount of damper force. The positions are numbered, from 1 to 4, by stamp marks on the adjuster wheel. As the adjuster wheel is turned, there is a positive click stop as each position is engaged. The settings are:

Position 1 (least damping force)
Position 4 (greatest damping force)

The amount of damping force must always correspond accurately with the spring tension selected. See the accompanying table. This enables correct use of the machine, both at high speed and with a passenger and/or heavy load added, whilst maintaining stability and a certain amount of comfort. Care must also be taken to ensure that the damping force rate, and spring tension settings are exactly the same on both rear suspension units, as again, instability will result from a failure to do this.
4   Having adjusted the rear suspension to the required settings for the general usage of the machine, the front suspension air-assistance rate, must now be checked. Re-adjusting either front or rear suspension alone is not recommended as the handling properties of the machine will again be impaired, possibly to a dangerous degree due to a serious imbalance occuring between front and rear suspension operations. See the accompanying table for the recommended front fork air pressure and rear suspension spring and damper settings.
5   The suspension units are sealed and there is no means of topping up or changing the damping fluid. If the damping fluid fails or if the unit leaks, renewal is necessary. If renewal is the only course available, the units must be treated, and replaced, as a matched pair.
6   On UK GS1000 EC and EN models the rear suspension is similar in operation to the front forks. The units retain oil damping with the addition of variable air-assistance. A valve is built-in to the top of each suspension unit enabling the exact pressure to be chosen by the operator. There are no specific recommendations laid down by Suzuki, as to the pressures suit-

able for differing riding conditions. The pressure should, therefore, be set to suit the individual preferences of the operator, taking, as a guide-line the pressure settings of the front forks. A minimum pressure of 2.0 kg/cm² (28.5 psi) and a maximum of approximately 2.1 kg/cm² (30 psi) is usual for normal solo riding. If a passenger or heavy load is carried regularly an increase in the rear suspension pressure is necessary. This could be up to 2.1 kg/cm² (31 psi). The maximum permissible pressure for a short period of time is 3.4 kg/cm² (48.5 psi). The front and rear suspension air pressure **must** be matched to ensure safe handling of the machine. The recommended pressures are as follows.

Front	Rear
0.80 – 0.98 kg/cm²	*2.0 – 2.1 kg/cm²*
(11.5 – 14 psi)	*(28.5 – 30 psi)*
0.98 – 1.1 kg/cm²	*2.1 – 2.2 kg/cm²*
(14 – 17 psi)	*(30 – 31 psi)*

7   The method of pumping up the rear units is, similar to that used on the front forks. A hand-pump, with suitable adaptor, should be connected to the valve at the top of each rear unit. Ensure the balance between the two units is correct; with not more than 0.1 kg/cm² (1.4 psi) difference.

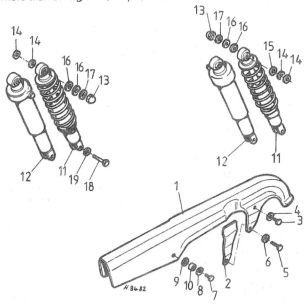

Fig. 4.5 Rear suspension units

1	Chain guard	11	Rear suspension unit – 2 off
2	Chain guard bracket	12	Rear suspension unit – 2 off
3	Bolt	13	Domed nut – 2 off
4	Washer	14	Inner washer – 4 off
5	Bolt	15	Right-hand inner washer
6	Spring washer	16	Outer washer – 4 off
7	Bolt	17	Outer washer – 2 off
8	Washer	18	Bolt – 2 off
9	Grommet	19	Washer – 2 off
10	Spacer		

Spring setting	Damper setting
I	1 or 2
II	2 or 3
III	3 or 4
IV	3 or 4
V	4

Fig. 4.6 Suspension unit adjustment balance table except UK GS1000 EC and EN models

Spring setting	Damper	Front fork air pressure
I	1	0.8 kg/cm² (11 psi)
I	2	0.8 ~ 0.9 kg/cm² (11 ~ 13 psi)
II	2	0.8 ~ 0.9 kg/cm² (11 ~ 13 psi)
II	3	0.8 ~ 0.9 kg/cm² (11 ~ 13 psi)
III	3	1.0 ~ 1.1 kg/cm² (14 ~ 16 psi)
III	4	1.0 ~ 1.1 kg/cm² (14 ~ 16 psi)
IV	3	1.0 ~ 1.1 kg/cm² (14 ~ 16 psi)
IV	4	1.0 ~ 1.1 kg/cm² (14 ~ 16 psi)
V	4	1.2 kg/cm² (17 psi)

Fig. 4.7 Front/rear suspension adjustment balance table except UK GS1000 EC and EN models

10.3 Prise up rubber boot at top of rear suspension unit and use screwdriver blade to facilitate damping adjustment

11 Centre stand: examination

1   The centre stand is retained on the underside of the frame by two bolts which serve as pivot shafts. A bush if fitted to each shaft. The pivot assemblies on centre stands are often neglected with regard to lubrication and this will eventually lead to wear. It is prudent to remove the pivot bushes from time to time and grease them thoroughly. This will prolong the effective life of the stand.
2   Check that the return spring is in good condition. A broken or weak spring may cause the stand to fall whilst the machine is being ridden, and catch in some obstacle, unseating the rider.

12 Prop stand: examination

1   The prop stand bolts to a lug attached to the rear of the left-hand lower frame tube. An extension spring ensures that the stand is retracted when the weight on the machine is taken off the stand.
2   Check that the pivot bolt is secure and that the extension spring is in good condition and not over-stretched. An accident is almost inevitable if the stand extends whilst the machine is on the move.

13 Footrests: examination and renovation

1   The front footrests on all models are of the bolt-on type, with fixed rubber pads. If they are bent in a spill or through the machine falling over, they can be removed and straightened in a vice whilst heated to a dull red with a blow lamp, or welding torch. The pillion footrests are hinged and therefore are less likely to become damaged than the front footrests. Each peg pivots on a clevis pin secured by a washer and split pin.

14 Brake pedal: examination

1   The rear brake pedal is secured by a single pinch bolt to the splined brake pivot shaft. In the event of damage, the pedal may be removed and treated similarly to a bent footrest, as described in the previous section.
2   The pedal may be pulled off the shaft after removing completely the pinch bolt. The outer end of the brake return spring, which shares the pivot shaft, should be displaced from the anchor peg on the frame, so that it may be removed at the same time as the pedal.

15 Dualseat: removal and replacement

1   The dualseat is attached to two lugs on the left side of the frame by two clevis pins secured with split pins. If it is necessary to remove the dualseat, withdraw the two split pins, take out the clevis pins, and the seat will lift off as a complete unit.

16 Speedometer and tachometer heads: removal and replacement

1   On the GS1000 models the speedometer and tachometer are mounted together on a single panel on top of the front forks. Each instrument does, however, have a separate holder which secures the instruments to the shared base plate. The separate holders are held to the instruments by two studs projecting from the base of each instrument, and then passing through the shared base plate and into the base cover, to be retained by nuts, on the underside of the cover. A separate plastic case and rubber sealing ring is fitted around the two instruments. The warning light console situated between the speedometer and tachometer, can be detached once the plastic instrument case has been removed. This case can be detached separately from the rest of the instruments, if required, once the trip-meter reset knob has been removed from the side of the speedometer unit. To detach the reset knob, remove the single retaining screw. The warning lamp console is retained by four screws around its periphery.
2   The instruments may be detached from the machine as a unit after disconnecting the drive cables and fuel gauge and warning light leads at the block connectors. After removing the base plate, which is retained by two large bolts, the bulb holders may be pulled from position. The foregoing applies to the GS1000 models C, HC, EC, and EN. On the L model, the mounting of the instruments is similar to that of the other models. Each instrument does, however, have a separate case instead of the one-piece unit used on the other models. The warning lamp console, fitted in the same position as on the other models, contains a series of lamps which light up individually to display, digitally, the gear position that has been selected. It is necessary, therefore, to disconnect the leads for this device before detaching the instrument unit base plate, on the L model.

3   With the small fairing attached to the top of the front forks on the S model, plus the addition of the extra instrumentation; the clock and oil temperature gauge, various differences are apparent during the dismantling of the main instruments.
4   Remove the instrument housing from around the speedometer and tachometer, after the trip-meter reset button, situated to the left of the speedometer in its own housing, has been freed by the removal of its retaining screw. Access to the reset button screw is from the underside of the unit, with the fairing removed. Also remove the screw which retains the adjusting button for the clock, in the instrument upper console. With the screw removed, detach the button. Lift away the instrument housing.
5   The speedometer and tachometer are mounted on separate holders, each retained to the shared base plate by four screws. The base plate is held to the top of the forks by four bolts. The forwardmost two bolts also serve to act as bracket retainers for the fairing. Detach the wiring leads from the speedometer and tachometer, to enable them to be removed as separate units, with the holder screws removed.
6   On all models, if either instrument fails to record, check the drive cable first, before suspecting the head. If the instrument gives a jerky response it is probably due to a dry cable, or one that is trapped or kinked.
7   The speedometer and tachometer heads cannot be repaired by the private owner, and if a defect occurs a new instrument has to be fitted. Remember that a speedometer in correct working order is required by law on a machine in the UK and also in many other countries.
8   Speedometer and tachometer cables are only supplied as a complete assembly. Make sure the cables are routed correctly through the clamps provided on the top fork yoke, brake branch pipe, and the frame.

17 Speedometer and tachometer drives: location and examination

1   The speedometer is driven from a gearbox fitted to the front wheel spindle on the left-hand side of the hub. Drive is transmitted through a dog plate fixed to the hub which engages with the drive gear in the gearbox.
2   Provided that the gearbox is repacked with grease from time to time, very little wear should be experienced. In the event of failure, the complete gearbox should be renewed. The tachometer drive is taken from the cylinder head cover, between number three and four cylinders. The drive is taken from the overhead camshaft by means of skew-cut pinions, and then by a flexible cable to the tachometer head. It is unlikely that the drive will give trouble during the normal service life of the machine, especially since it is fully enclosed and effectively lubricated.

18 Cleaning the machine

1   After removing all surface dirt with a rag or sponge which is washed frequently in clean water, the machine should be allowed to dry thoroughly. Application of car polish or wax to the cycle parts will give a good finish, particularly if the machine receives this attention at regular intervals.
2   The plated parts should require only a wipe with a damp rag, but if they are badly corroded, as may occur during the winter when the roads are salted, it is permissible to use one of the proprietary chrome cleaners. These often have an oily base which will help to prevent corrosion from recurring.

3   If the engine parts are particularly oily, use a cleaning compound such as Gunk or Jizer. Apply the compound whilst the parts are dry and work it in with a brush so that it has an opportunity to penetrate and soak into the film of oil and grease. Finish off by washing down liberally, taking care that water does not enter the carburettors, air cleaners or the electrics.

4   If possible, the machine should be wiped down immediately after it has been used in the wet, so that it is not garaged under damp conditions which will promote rusting. Make sure that the chain is dried and lightly greased or coated sparingly with ordinary heavyweight oil. Do not use one of the proprietary aerosol spray lubricants as these may be harmful to the O-rings used in the construction of the chain. This will prevent water from entering the rollers and so causing harshness with an accompanying rapid rate of wear. Remember there is less chance of water entering the control cables and causing stiffness if they are lubricated regularly as described in the Routine Maintenance Section.

19 Fault diagnosis: frame and forks

Symptom	Cause	Remedy
Machine is unduly sensitive to road conditions	Forks and/or rear suspension units have defective damping	Check oil level/air pressure in front forks. Check rear suspension units for incorrect damping/spring setting. Adjust as necessary.
Machine tends to roll at low speeds	Steering head bearings overtight or damaged	Slacken bearing adjustment. If no improvement, dismantle and inspect bearings.
Machine tends to wander, steering is imprecise	Worn swinging arm bearings	Check and if necessary renew bearings.
Fork action stiff	Fork legs have twisted in yokes or have been drawn together at lower ends	Slacken off spindle nut clamps, pinch bolts in fork yokes and fork top nuts. Pump forks several times before re-tightening from bottom
Forks judder when front brake is applied	Worn fork legs and stanchions Steering head bearings too slack	Renew one or both items. Readjust to take up play.
Wheels out of alignment	Frame distorted as a result of accident damage	Check frame alignment after stripping out. If bent, specialist repair is necessary.

Chapter 5 Wheels, brakes and tyres

Contents

Specifications

Tyres

Front	3.25V x 19 – 4PR (S model)
	3.50V x 19 – 4PR (All models except S)
Rear	4.00V x 18 – 4PR (C, N [USA], S models)
	4.50V x 17 – 4PR (HC, EC, EN, L models)

Tyre pressures

	Solo	Pillion
Front	25 psi (1.75 kg cm^2)	28 psi (2.00 kg cm^2)
Rear	28 psi (2.00 kg cm^2)	32 psi (2.25 kg cm^2)

For continuous high-speed riding, the pressures should be increased to:

	Solo	Pillion
Front	28 psi (2.00 kg cm^2)	32 psi (2.25 kg cm^2)
Rear	36 psi (2.50 kg cm^2)	40 psi (2.80 kg cm^2)

Minimum recommended tread depth

Front	1.6 mm (0.06 in)
Rear	2.0 mm (0.08 in)

Wheel rim runout (maximum)

	Axial	Radial
Front	2.0 mm (0.08 in)	2.0 mm (0.08 in)
Rear	2.0 mm (0.08 in)	2.0 mm (0.08 in)

Brakes

Front:

C and N models	Single 11.5 in (292 mm) hydraulically operated disc
HC, EC, EN, L, S models	Twin 10.8 in (274 mm) hydraulically operated discs
Rear (all models)	Single 10.8 in (274 mm) hydraulically operated disc

Disc thickness

C and N models

Front	6.7 mm (0.264 in)
Service limit	6.0 mm (0.236 in)
Rear	6.7 mm (0.264 in)
Service limit	6.0 mm (0.236 in)

HC, EC, EN, L, S models

Front	5.0 mm (0.20 in)
Service limit	4.5 mm (0.18 in)
Rear	As C and N models

Disc runout (maximum)

0.3 mm (0.011 in)

Brake fluid specifications

DOT 3 (USA), SAE J1703 (UK)

Main torque wrench settings

Front spindle:	
Clamp nut (8 mm)	1.5 – 2.5 kgf m (10.8 – 18.0 lbf ft)
Spindle nut (12 mm)	3.6 – 5.2 kgf m (26.0 – 37.6 lbf ft)
Front caliper mounting bolt (10 mm)	2.5 – 4.0 kgf m (18.8 – 29.0 lbf ft)
Rear spindle nut (18 mm)	8.5 – 11.5 kgf m (61.5 – 83.0 lbf ft)
Rear caliper mounting bolt (10 mm)	2.0 – 3.0 kgf m (14.5 – 21.7 lbf ft)

1 General description

All models within the GS1000 range are fitted with a 19 inch diameter front wheel. All the models fitted with aluminium alloy wheels, excepting the S, have 17 inch diameter rear wheels. The GS1000 S model is fitted with an 18 inch diameter rear wheel.

The front tyre fitted to all models in the range has a 3.50 inch section except for the S model which has a 3.25 inch section. On the models with the 17 inch diameter rear wheel, a large, 4.50 inch, section tyre is the standard fitting. The GS1000 S has a 4.00 inch section rear tyre fitted.

The C and HC models as sold in the USA are fitted with traditional wheels, having chromed steel rims, laced to an alloy hub by butted wire spokes. All the models marketed in the UK, and the EC, EN, and L models, plus the 'limited edition' number of S models, in the USA, are fitted with cast alloy wheels.

The GS1000 C and HC (USA versions) are equipped with a single hydraulically-operated disc brake on both the front and rear wheels. All other models sold in America and the UK are fitted with twin front disc brakes, and a single rear disc brake.

2 Front wheel: examination and renovation (wire spoked wheels)

1   Place the machine on the centre stand so that the front wheel is raised clear of the ground. Spin the wheel and check the rim alignment. Small irregularities can be corrected by tightening the spokes in the affected area although a certain amount of experience is necessary to prevent over-correction. Any flats in the wheel rim will be evident at the same time. These are more difficult to remove and in most cases it will be necessary to have the wheel rebuilt on a new rim. Apart from the effect on stability, a flat will expose the tyre bead and walls to greater risk of damage if the machine is run with a deformed wheel.

2   Check for loose and broken spokes. Tapping the spokes is the best guide to tension. A loose spoke will produce a quite different sound and should be tightened by turning the nipple in an anticlockwise direction. Always check for run out by spinning the wheel again. If the spokes have to be tightened by an excessive amount, it is advisable to remove the tyre and tube as detailed in Section 20 of this Chapter. This will enable the protruding ends of the spokes to be ground off, thus preventing them from chafing the inner tube and causing punctures.

3 Front wheel: examination and renovation (cast wheels)

1   With the cast alloy wheels, a careful check must be made for cracks and chipping, particularly at the spoke bases and the edge of the rim. As a general rule a damaged cast wheel must be renewed, as cracks will cause stress points which may lead to sudden failure under heavy load. Small nicks may be radiused carefully with a fine file and emery paper (No. 600 – No. 1000) to relieve the stress. If there is any doubt as to the condition of the wheel, advice should be sought from a Suzuki repair specialist.

2   Each wheel is covered with a coating of lacquer, to prevent corrosion. If damage occurs to the wheel, and the lacquer finish is penetrated, the bared aluminium alloy will soon start to corrode. A whitish grey oxide will form over the damaged area, which in itself is a protective coating. This deposit, however, should be removed carefully whenever possible, and a new protective coating of lacquer applied.

3   Check the lateral run-out at the rim by spinning the wheel and placing a fixed pointer close to the rim edge. If the maximum run-out is greater than 2.0 mm (0.08 in), Suzuki recommend that the wheel be renewed. This is, however, probably slightly over-cautious; a run-out somewhat greater than this can probably be accommodated without noticeable effect on steering. No means is available for straightening a warped wheel without resorting to the expense of having the wheel skimmed on all faces. If warpage has occured as the result of impact during an accident, the safest measure is to renew the wheel. Worn wheel bearings may cause rim run-out. These should be renewed as described in Section 9 of this Chapter.

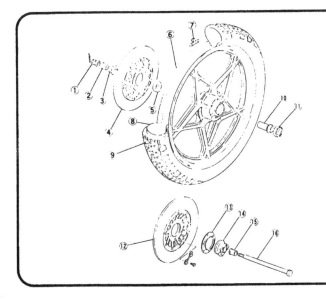

Fig. 5.1 Front wheel assembly

1   Castellated nut
2   Right-hand spacer
3   Bearing right-hand spacer
4   Front brake disc
5   Right-hand bearing
6   Front wheel
7   Balance weight AR
8   Inner tube
9   Tyre
10   Spacer
11   Left-hand bearing
12   Front brake disc
13   Dust cover
14   Speedometer gearbox assembly
15   Left-hand spacer
16   Front wheel spindle

4 Front wheel: removal and replacement

1   Place the machine on the centre stand so that it is resting securely on firm ground with the front wheel well clear of the ground. If necessary, place wooden blocks below the crankcase to raise the wheel.

2   Where two brake calipers are utilised, one must be detached from the fork leg to allow clearance for the wide section tyre. Remove the caliper as a complete unit – without disconnecting the hydraulic fluid hose – by unscrewing the two bolts which pass through the caliper support bracket and fork leg.

3   Disconnect the speedometer at the gearbox by unscrewing the knurled ring. Pull the cable through the guide clip. Displace the split pin from the wheel spindle nut and slacken the nut slightly. The wheel may be removed either by detaching the two spindle clamps or by slackening the clamp bolts and withdrawing the spindle. In the latter case the speedometer gearbox and wheel spacer will fall free as the wheel is lowered from place.

4   When refitting the wheel into the forks, ensure that the speedometer gearbox is fitted with the embossed arrow-mark pointing upwards. Tighten the wheel spindle nut fully before tightening the two spindle clamps, and do not omit the split pin. The two nuts holding each spindle clamp should be tightened down evenly so that the gap between the clamp and fork leg is equal either side of the wheel spindle.

4.3a Disconnect the speedometer cable at the drive gearbox

4.3b Withdraw the split pin, or R pin, from the wheel spindle nut and ...

4.3c ... slacken the nut

4.3d Slacken the clamp bolts and ...

4.3e ... withdraw the spindle

4.3f Speedometer gearbox will fall free as wheel is removed

5 Front brake assembly: examination and brake pad renewal

1   Check the front brake master cylinder, hoses and caliper units for signs of leakage. Pay particular attention to the condition of the hoses, which should be renewed without question if there are signs of cracking, splitting or other exterior damage. Check the hydraulic fluid level by referring to the upper and lower level visible on the exterior of the transparent reservoir body.

2   Replenish the reservoir after removing the cap on the brake fluid reservoir and lifting out the diaphragm plate. The condition of the fluid can be checked at the same time. Checking the fluid level is one of the maintenance tasks which should **never be neglected.** If the fluid is below the lower level mark, brake fluid of the correct specification must be added. **Never** use engine oil or any fluid other than that recommended. Other fluids have unsatisfactory characteristics and will rapidly destroy the seals. The fluid level is unlikely to fall other than a small amount, unless leakage has occured somewhere in the system. If a rapid change of level is noted a careful check for leaks should be made before the machine is used again.

3   The two sets of brake pads should be inspected for wear. Each has a red groove, which marks the wear limit of the friction material. When this limit is reached, both pads in the set must be renewed, even if only one has reached the wear mark. In normal use both sets of pads will wear at the same rate and therefore both sets must be renewed.

4   If the brake action becomes spongy, or if any part of the hydraulic system is dismantled (such as when a hose has been renewed) it is necessary to bleed the system in order to remove all traces of air. Follow the procedure in Section 7 of this Chapter.

5   To gain access to the pads for renewal, the caliper assembly must be detached from the front fork. Removal of the wheel is not required, nor is separation of the caliper from the hydraulic hose. Remove the two bolts which pass through the fork leg into the caliper support bracket and lift the complete caliper unit upwards, off the disc.

6   Remove the single screw and the convolute backing plate from the inner side of the caliper unit. The inner pad is now free and may be displaced towards the centre of the caliper and lifted out. The outer pad which abuts against the caliper piston is not retained positively and may be lifted out.

7   Refit the new pads and replace the caliper by reversing the dismantling procedure. The caliper piston should be pushed inwards slightly so that there is sufficient clearance between the brake pads to allow the caliper to fit over the disc. It is recommended that the outer periphery of the outer (piston) pad is lightly coated with disc brake assembly grease (silicon grease). Use the grease sparingly and ensure that grease **Does Not** come in contact with the friction surface of the pad.

5.5 Caliper is held to the fork by two chromed bolts

5.6a Remove screw (arrowed) and backing plate and ...

5.6b ... displace inner pad towards piston pad

5.6c The piston pad lifts out

6 Front brake caliper: examination and overhaul

1   Select a suitable receptacle into which may be drained the hydraulic fluid. Remove the banjo bolt holding the hydraulic hose at the caliper and allow the fluid to drain. Repeat this operation with the second caliper unit (where fitted). Take great care not to allow hydraulic fluid to spill onto paintwork; it is a very effective paint stripper. Hydraulic fluid will also damage rubber and plastic components.

2   Remove the caliper from the fork leg and displace the brake pads as described in the preceding Section. Where two calipers are fitted they should be dismantled and reassembled individually, to prevent the accidental interchange of components.

3   Remove the two Allen bolts which pass through the caliper body, and separate the body from the caliper support bracket. Prise out the piston boot, using a small screwdriver, taking care not to scratch the surface of the cylinder bore. The piston can be displaced most easily by apply an air jet to the hydraulic fluid feed orifice. Be prepared to catch the piston as it falls free. Displace the annular piston seal from the cylinder bore groove.

4   Clean the caliper components thoroughly in a fine solvent, or in hydraulic brake fluid. **CAUTION**: Never use petrol for cleaning hydraulic brake parts otherwise the rubber components will be damaged. Discard all the rubber components as a matter of course. The replacement cost is relatively small and does not warrant re-use of components vital to safety. Check the piston and caliper cylinder bore for scoring, rusting or pitting. If any of these defects are evident it is unlikely that a good fluid seal can be maintained and for this reason the components should be renewed.

5   To assemble the caliper, reverse the removal procedure. When assembling pay attention to the following points. Apply Suzuki caliper grease (high heat resistance) to the caliper spindles and note the two O-rings on the forwardmost of the two spindles. These should be renewed upon reassembly. Apply a generous amount of brake fluid to the inner surface of the cylinder and to the periphery of the piston, then assemble. Do not assemble the piston with it inclined or twisted. When installing the piston push it slowly into the cylinder while taking care not to damage the piston seal. Apply Suzuki brake pad grease around the periphery of the moving pad. Bleed the brake after refilling the reservoir with new hydraulic brake fluid, then check for leakage while applying the brake lever tightly. After a test run, check the pads and brake disc.

6   Note that any work on the hydraulic system must be undertaken under ultra-clean conditions. Particles of dirt will score the working parts and cause early failure of the system.

6.3a Removal of the two Allen-head caliper spindles allows ...

6.3b ... separation of the body from the support bracket

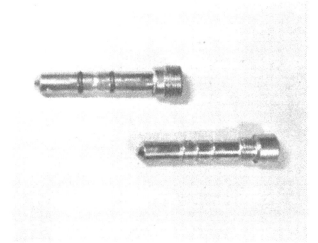

6.5a Note the two O-rings on the forward-most caliper spindle

6.5b Refit and securely tighten screw on inner pad

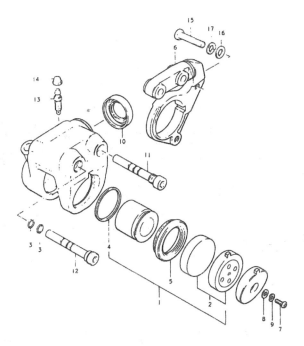

Fig. 5.2 Front brake caliper

1	Piston/pad set	10	Boot
2	Disc pad – 2 off	11	Spindle
3	O-ring – 2 off	12	Spindle
4	Piston seal	13	Bleed nipple
5	Boot	14	Dust cap
6	Caliper bracket	15	Bolt – 2 off
7	Screw	16	Washer – 2 off
8	Washer	17	Spring washer – 2 off
9	Spring washer		

7 Front disc brake master cylinder: examination and renovation

1   The master cylinder and hydraulic reservoir take the form of a combined unit mounted on the right-hand side of the handlebars, to which the front brake lever is attached. The master cylinder is actuated by the front brake lever, and applies hydraulic pressure through the system to operate the front brake when the handlebar lever is manipulated. The master cylinder pressurises the hydraulic fluid in the brake pipe which, being incompressible, causes the piston to move in the caliper unit and apply the friction pads to the brake. If the master cylinder seal leaks, hydraulic pressure will be lost and the braking action rendered much less effective.

2   Before the master cylinder can be removed the system must be drained. Place a clean container below the caliper unit and attach a plastic tube from the bleed screw on top of the caliper unit to the container. Open the bleed screw one complete turn and drain the system by operating the brake lever until the master cylinder reservoir is empty. Close the bleed screw and remove the pipe.

3   Remove the front brake stop lamp switch from the master cylinder (USA and Canadian specification). Unscrew the union bolt and disconnect the connection between the brake hose and the master cylinder. Unscrew the two master cylinder fastening bolts and remove the master cylinder body from the handlebar. Empty any surplus fluid from the reservoir.

4   Remove the brake lever from the body, remove the boot stopper (taking care not to damage the boot) and then remove the boot. Remove the circlip that was hidden by the boot, the piston, primary cup, spring and check valve. Place the parts in a clean container and wash them in new brake fluid. Examine the cylinder bore and piston for scoring. Renew if scored. Check also the brake lever for pivot wear, cracks or fractures and the hose union threads and brake pipe threads for cracks or other signs of deterioration.

5   When assembling the master cylinder follow the removal procedure in reverse order. Pay particular attention to the following points: Make sure the primary cup is fitted the correct way round. Renew the split pin of the brake lever pivot nut and fit it securely. Mount the master cylinder to the handlebar so the gap between it is 2 mm (0.08 in) and the reservoir is horizontal when the motorcycle is on the centre stand with the steering in the straight ahead direction. Fill with fresh fluid and bleed the system. Be sure to check the brake reservoir by removing the reservoir cap. If the level is below the ring mark inside the reservoir, refill the level with the prescribed brake fluid.

6   The component parts of the master cylinder assembly and the caliper assembly may wear or deteriorate in function over a long period of use. It is however, generally difficult to foresee how long each component will work with proper efficiency and from a safety point of view it is best to change all the expendable parts every two years on a machine that has covered a normal mileage.

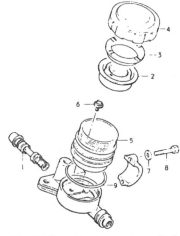

Fig. 5.3 Front brake master cylinder

1	Piston/seal assembly	6	Screw – 2 off
2	Diaphragm	7	Washer – 2 off
3	Seal	8	Bolt – 2 off
4	Cap	9	O-ring
5	Reservoir		

8 Bleeding the hydraulic brake system

1   If the hydraulic system has to be drained and refilled, if the brake lever travel becomes excessive or the lever operates with a soft or spongy feeling, the brakes must be bled to expel air from the system. The procedure for bleeding the hydraulic brake is best carried out by two persons. The following procedure applies equally to front or rear brakes.

2   First check the fluid level in the reservoir and top up with fresh fluid.

3   Keep the reservoir at least half full of fluid during the bleeding procedure.

4   Screw the cap on to the reservoir to prevent a spout of fluid or the entry of dust into the system. Place a clean glass jar below the caliper bleed screw and attach a clear plastic pipe from the caliper bleed screw to the container. Place some clean hydraulic fluid in the jar so that the pipe is always immersed below the surface of the fluid.

5   Unscrew the bleed screw one half turn and squeeze the brake lever as far as it will go but do not release it until the bleeder valve is closed again. Repeat the operation a few times until no more air bubbles come from the plastic tube.

6   Keep topping up the reservoir with new fluid. When all the bubbles disappear, close the bleeder valve. Remove the plastic tube and install the bleeder valve dust cap. Check the fluid level in the reservoir, after the bleeding operation has been completed. On machines fitted with twin front brakes the procedure should be repeated on the second caliper.

7   Reinstall the diaphragm and tighten the reservoir cap securely. Do not use the brake fluid drained from system, since it will contain minute air bubbles.

8   Never use any fluid other than that recommended. Oil must not be used under any circumstances.

9 Front wheel bearings: examination and replacement

1   Access to the front wheel bearings can be made after removal of the speedometer gearbox and spindle spacer.

2   The wheel bearings can be drifted out of position, using a suitable drift. Support the wheel so the exit of the bearing is not obstructed. When the first bearing has been removed the spacer that lies between the two bearings can be removed. Insert the drift and drive out the opposite bearing.

3   Remove all the old grease from bearings and hub. Wash the bearings in petrol and dry them thoroughly. Check the bearings for roughness by spinning them whilst holding the inner track

with one hand and rotating the outer track with the other. If there is the slightest sign of roughness renew them. Bearings fitted with seals both sides cannot be cleaned or regreased. They should be renewed as a matter of course if not in first class condition.

4   Before driving bearings back into the hub, pack the hub with new grease and also grease the bearings. Use the same double diameter drift to place them in position. Refit any oil seals or dust covers which have been displaced.

10 Removing and replacing the disc

1   It is unlikely that the brake discs will require attention unless bad scoring has developed or the discs have warped. To detach the discs first remove the wheel as described in Section 3 of this Chapter. Each disc is retained by six bolts screwed into the hub, which are linked in pairs by tab washers. Bend down the ears of the tab washers and remove the bolts. The discs can then be eased off the hub bosses.

2   Replace the discs by reversing the dismantling procedure. Ensure that the twelve nuts are tightened fully and that the tab washer ears are bent up against the bolt head flats.

9.2a With the bearing removed, the centre spacer is free to be pulled out

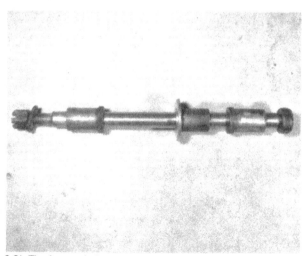

9.2b The front spindle with both spindle spacers and the flanged bearing spacer

9.2c Refit axle with spacers in this position

9.4 Grease the bearings before refitting with drift

10.1 Disc is retained by bolts secured by locking plates

11 Rear wheel: examination, removal and renovation

1   Place the machine on the centre stand so that the rear wheel is raised clear of the ground. Follow the procedures laid down in Section 2 & 3 of this Chapter, relating to the two differing types of wheel fitted to the GS1000 range, and check for rim alignment, rim damage, loose spokes etc.

2   Removal of the rear wheel will be accomplished far more easily if the plastic chainguard is detached first. Remove the two retaining bolts and lift the chain-guard away.

3   Although not strictly necessary, it is advised that both silencers are removed to improve access. It is however, not possible to withdraw the rear spindle without first removing the left-hand side silencer. Each silencer is retained by a single bolt passing through a lug on the underside of the silencer body. It will be necessary also to slacken the two clamps at the silencer/exhaust pipe joints.

4   Unscrew the bolt passing vertically through each chain adjuster block. The right-hand bolt also secures the hydraulic hose guide clamp. Remove the split pin and nut securing the torque arm rod to the top of the caliper support bracket. Detach the caliper from the support bracket by removing the two mounting bolts. Position the caliper and its attached hose so that it does not become damaged and does not obstruct further dismantling. Suspending the unit from the rear right-hand indicator stalk should serve this purpose satisfactorily.

5   After loosening the locknuts, unscrew the two wheel adjuster bolts so that the adjuster brackets can be swung downwards below the fork ends. Remove the R-clip securing the castellated spindle nut and slacken the nut. If the right-hand silencer has not been detached, there may be insufficient clearance to allow the nut to fully run off the thread on the end of the spindle. The spindle should, therefore, be partially withdrawn, to allow the nut to be removed, and then reinserted.

6   It is now necessary to obtain sufficient chain slack to facilitate removal of the wheel. Push the wheel forwards fully and lift the chain off the sprocket, pulling to the outside of the fork end and then forwards to clear the wheel spindle head. If there is insufficient slack in the chain, as would be the case with a new or recently adjusted chain, then it must be detached from the machine complete with the primary drive sprocket. Due to the one-piece construction of the chain, there is no spring link, it cannot be 'split'. No attempt should be made to separate the chain. See Chapter 1, Section 5, paragraph 10.

7   Refit the wheel by reversing the dismantling procedure. Ensure that the torque arm is secure and that the securing split pins are fitted. Likewise, do not omit the wheel spindle nut securing R-pin. Before tightening the spindle nut and chain adjuster locknuts, the final drive chain should be adjusted. Adjust the chain so that there is 20 mm (0.8 in) of up and down play measured in the centre of the chain lower run. Refer to the index marks on the fork ends when tightening the adjuster bolts, to ensure that wheel alignment is maintained. Do not omit to tighten the chain adjuster locknuts when tension is correct.

12 Rear disc brake: examination and pad renewal

1   In general, remarks concerning the front disc brake as described in Section 5 of this Chapter apply equally to the rear disc brake. The rear brake master cylinder/reservoir unit is fitted to the right-hand side of the machine behind the frame side cover.

2   To inspect the brake pads for wear, prise off the inspection cap fitted to the top of the caliper. Each pad is stepped slightly, the step nearer the backing plate being painted red. If either pad is worn down to the red line, the pads must be renewed as a set.

3   Pad removal may take place without displacing the caliper unit or the wheel. Pull out the stop pin which passes through each of the two mounting pins. Displace one mounting pin and remove the two hair springs. Push out the final pin and lift each pad out individually, removing the outer pad first.

4   Install new pads by reversing the dismantling procedure. It will probably be necessary to push back each piston to give the required clearance between the piston and the disc; the new pads will be considerably thicker. The metal shim fitted to the piston side of each pad must be positioned with the punched arrow mark pointing in the direction of wheel travel, i.e. the point of the arrow faces forwards.

13 Rear disc brake caliper: examination and overhaul

1   Unlike the front brake caliper, which has only one piston. The rear brake caliper has two pistons and two moving brake pads. The general procedure however for dismantling and overhaul is similar to that described for the front brake in Section 6 of this Chapter.

2   After reassembling the caliper, the hydraulic circuit should be bled of all air as described in Section 8. The inboard side of the caliper should be bled first using the inner bleed nipple.

11.2a Remove the lower rear bolt and ...

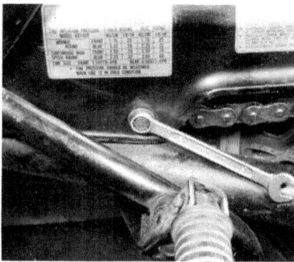

11.2b ... the forward bolt to remove chainguard

11.4a The right-hand chain adjuster block bolt secures brake hose clamp

11.4b Remove the split pin and nut securing the torque arm

11.4c The support bracket with caliper unit removed

11.7a On refitting rear wheel, check spindle nut and R-clip and ...

11.7b ... torque arm nut and split pin are fitted securely

11.7c Refit and tighten rear caliper mounting bolts

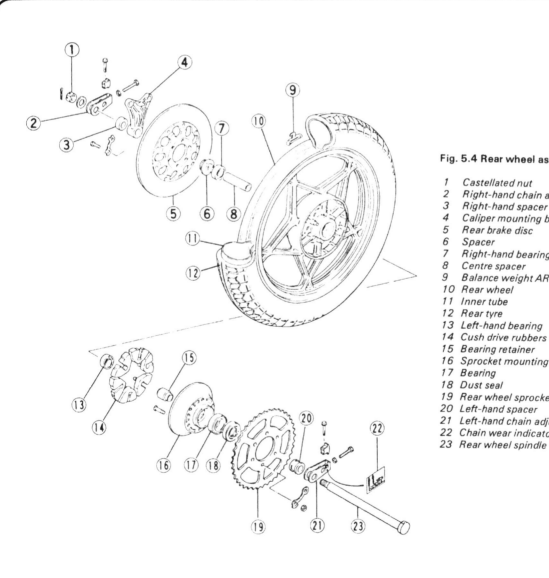

Fig. 5.4 Rear wheel assembly

1   Castellated nut
2   Right-hand chain adjuster
3   Right-hand spacer
4   Caliper mounting bracket
5   Rear brake disc
6   Spacer
7   Right-hand bearing
8   Centre spacer
9   Balance weight AR
10 Rear wheel
11 Inner tube
12 Rear tyre
13 Left-hand bearing
14 Cush drive rubbers
15 Bearing retainer
16 Sprocket mounting drum
17 Bearing
18 Dust seal
19 Rear wheel sprocket
20 Left-hand spacer
21 Left-hand chain adjuster
22 Chain wear indicator label
23 Rear wheel spindle

12.2 Detach caliper inspection cap to check pad condition

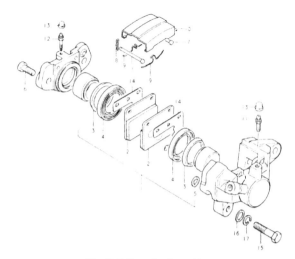

Fig. 5.5 Rear brake caliper

1   Piston/pad set
2   Pad – 2 off
3   Piston seal – 2 off
4   Boot – 2 off
5   O-ring
6   Bolt – 2 off
7   Pin – 2 off
8   R-pin – 2 off
9   Spring – 2 off
10   Cover
11   Bleed nipple
12   Bleed nipple
13   Dust cap – 2 off
14   Anti-squeal shim – 2 off
15   Bolt – 2 off
16   Washer – 2 off
17   Spring washer – 2 off

12.3 Remove the mounting pin stop pins, to enable the mounting pins to pull out

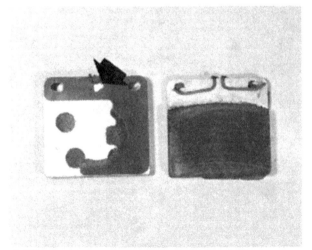

12.4 The two pads, note arrow punched in metal shim (arrowed)

14 Rear brake master cylinder: removal and overhaul

1   In common with the unit fitted to the front brake system, the rear brake master cylinder and fluid reservoir are a combined unit. The master cylinder is fitted inboard of the rear frame right-hand triangulation and is operated via a pushrod from the foot brake pedal.

2   To detach the master cylinder, remove the split pin and clevis pin from the lower end of the operating pushrod. Disconnect the hydraulic hose at the master cylinder by removing the banjo bolt, and allow the hose fluid to drain into a suitable container. The master cylinder is secured to the frame by two bolts passing through lugs on the master cylinder body.

3   Remove the reservoir cap and diaphragm and allow the fluid to drain. Pull the gaiter along the pushrod and remove the circlip to allow detachment of the pushrod assembly. Using a wood dowel inserted through the hydraulic fluid inlet orifice push out the piston/seal assembly.

4   Refer to the procedure given in Section 7 for inspection and reassembly details.

5   As an alternative method of removal, should the master cylinder not require overhauling, the complete unit of rear caliper, hose and master cylinder can be detached as one assembly. Having detached the master cylinder (see Section 14, paragraph 2 of this Chapter), release the brake fluid hose from its retaining clips on the right-hand swinging arm, and then remove the caliper (see Section 11, paragraph 4, of this Chapter). With some wriggling of the components around the frame tubes, the complete assembly can be removed.

Tyre changing sequence - tubed tyres

 Deflate tyre. After pushing tyre beads away from rim flanges push tyre bead into well of rim at point opposite valve. Insert tyre lever adjacent to valve and work bead over edge of rim.

Use two levers to work bead over edge of rim. Note use of rim protectors

 Remove inner tube from tyre

When first bead is clear, remove tyre as shown

 When fitting, partially inflate inner tube and insert in tyre

Work first bead over rim and feed valve through hole in rim. Partially screw on retaining nut to hold valve in place.

 Check that inner tube is positioned correctly and work second bead over rim using tyre levers. Start at a point opposite valve.

Work final area of bead over rim whilst pushing valve inwards to ensure that inner tube is not trapped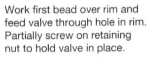

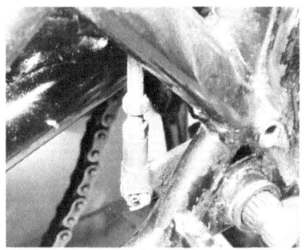

14.2a Detach the split pin and clevis pin from lower end of pushrod and ...

14.2b ... remove two bolts that retain master cylinder

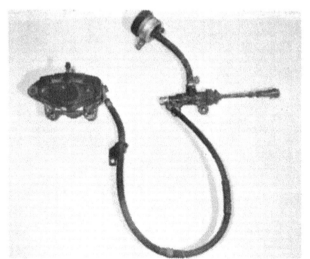

14.5 The complete rear brake assembly, removed as a unit

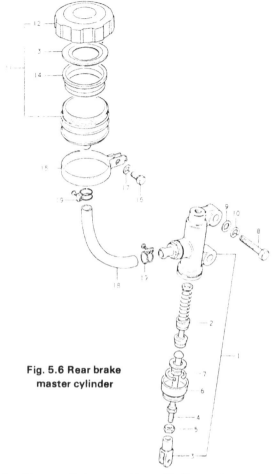

Fig. 5.6 Rear brake master cylinder

1	Master cylinder assembly	11	Reservoir assembly
2	Piston/seal assembly	12	Cap
3	Clevis fork	13	Seal
4	Pushrod	14	Diaphragm
5	Locknut	15	Clamp
6	Boot	16	Bolt
7	Circlip	17	Spring washer
8	Bolt – 2 off	18	Hose
9	Washer – 2 off	19	Clip – 2 off
10	Spring washer – 2 off		

15 Rear wheel bearings: removal and replacement

1   The rear wheel assembly has three journal ball bearings. One bearing lies each side of the wheel hub and the third bearing is fitted in the cush drive assembly to which is attached the sprocket. The rear wheel must be removed completely before the cush drive hub can be lifted from the wheel hub.

2   Drift the wheel bearings from position using the same method as described for the front wheel. Before the cush drive bearing is tapped out, the hollow spindle should be removed. It is not necessary to remove the sprocket. The cush drive hub bearing oil seal may be drifted out at the same time as the bearing.

16 Rear cush drive: examination and renovation

1   The cush drive assembly is contained in the left-hand side of the wheel hub. It takes the form of six triangular rubber pads incorporating slots, that fit within the vanes of the hub. A

heavily ribbed plate bolted to the rear sprocket engages with the slots to form a shock absorber which permits the sprocket to move within certain limits. This absorbs any surge or roughness in the transmission. The rubbers should be renewed when movement of the sprocket indicates bad compaction of the rubbers or if they commence to break up.

17 Rear wheel sprocket: examination and replacement

1   The rear wheel sprocket is held to the cush drive hub by six bolts locked by three tab washers. To remove the sprocket, bend back the locking tabs and undo the bolts. It must be noted that removal of the sprocket necessitates the removal of the hub from the wheel. If the nuts on the retaining bolts are attempted to be removed before the hub is separated from the wheel, the bolts will simply turn freely inside the hub. Similarly, during reassembly the bolts must be held secure from the inside during the refitting and tightening of the nuts and tab washers which retain the sprocket.

2   The sprocket need only be renewed if the teeth are worn or chipped. Although becoming an extremely expensive ideal, it is always a good policy to change both sprockets, and the chain, at the same time. If all three components are not replaced simultaneously, rapid wear will result on the new component(s).

18 Primary drive sprocket: examination and replacement

1   The primary drive sprocket is held to the splines on the gearbox output shaft by a large nut and lock washer. To remove the sprocket, bend down the locking tabs of the washer and undo and remove the bolt. This must be done with the chain still attached to the sprocket, and the two components must be drawn off the output shaft simultaneously, and then separated. To hold the rear wheel secure when undoing the sprocket retaining nut, instruct an assistant to apply the rear brake firmly.

2   The primary drive sprocket on the GS1000 range is of an unusual design, in that it incorporates noise-reducing neoprene dampers. These are fitted to each side of the sprocket, supported by metal flanges, and retained by three countersunk screws. Make a visual examination of the condition of the neoprene dampers; note whether there are signs of perishing or splitting. The condition of these dampers is not vital to the operation of the machine, but keeping them in good condition is conducive to a quieter operating machine. Renewal of the complete sprocket is necessary if the dampers are completely deteriorated. Other than for this reason, the sOrocket need only be replaced if wear or damage has occured, with the proviso stated in Section 16, paragraph 2, of this Chapter.

15.2a Drift wheel bearings out of position and ...

15.2b ... the hub bearing oil seal

16.1a The cush drive assembly (general view)

16.1b Examine the rubber pads for wear such as scoring

17.1a Cush drive hub is a push fit in rubber inserts

17.1b Six bolts and three locking tab washers retain the cush drive hub to the sprocket

17.1c Hub must be removed from wheel to undo retaining bolts

17.1d Do not omit spacer when replacing hub

18.2a The damping rubbers are retained by three screws and ...

18.2b ... the sprocket can be dismantled for inspection and/or replacement

19 Final drive chain : examination and lubrication

1   As the final drive chain is fully exposed on all models it requires lubrication and adjustment at regular intervals. To adjust the chain, take out the split pin from the rear wheel spindle and slacken the spindle nut. Undo the locknut on the chain adjusters and turn the adjuster bolts inwards to tighten the chain. Marks on the adjusters must be in line with identical marks on the frame fork to align the rear wheel correctly. A final check can be made by laying a straight wooden plank alongside the wheels, each side in turn. Chain tension is correct if there is 20 mm (0.8 in) of slack in the middle of the chain run between the two sprockets.

2   As a guide to determining the condition of the chain whilst it is still in position on the machine, Suzuki fit small coloured stickers to the rearmost part of the chain adjuster brackets. If the first indicator mark (red) aligns with the end of the swinging arm, then the chain must be replaced, as it is beyond its useful life. The second mark (green) indicates the position of the swinging arm, with a new chain fitted, in relation to the chain adjuster brackets.

3   Do not run the chain too tight to try to compensate for wear as it will absorb a surprising amount of engine power. Also it can damage the gearbox and rear wheel bearings. A chain that is run exceptionally slack is equally undesirable. Never run the chain with slack in the region of 50 mm (2.0 in). A chain that is run exceptionally slack will cause excessive wear to the sprockets and will render the machine a danger to ride. The results of an abnormally slack chain jumping off the sprockets, particularly if the occurence happened at the high speeds this machine is capable of, are potentially catastrophic. The resultant damage, to both the machine and the operator, would probably be expensive in the machines case, and possibly permanent in the case of the unfortunate operator. Even if the chain does not actually leave the sprockets, it may come into contact with the outer gearbox wall of the engine casing. The neutral indicator (and gear position indicator on L models) are situated on this side of the gearbox wall. If the chain strikes the switch(s), the switch body may break, causing oil loss from the gearbox. This may affect the continued lubrication of the gearbox components.

4   The chain fitted to the Suzuki GS1000 models is of unusual design in that when the chain is lubricated on initial assembly, the lubricant is sealed in for the life of the components by O-rings placed on each end of the rollers. Although the internal bearing surfaces of the chain are permanently lubricated, the outside of the rollers and the sprockets with which they mesh are not. Lubricant should be applied at intervals of approximately 1000 kms (600 miles) to prevent wear of the sprockets and chain rollers. If the prevailing weather or riding conditions are particularly arduous, increase the frequency of the lubrication. Use either a high melting point grease or a heavy-weight (i.e. SAE 40) engine oil for correct lubrication. The chain should *not* be immersed in a molten lubricant as is usual practice with chains, because the O-rings will suffer damage caused by the heat. Nor should one of the proprietary aerosol lubricants or any solvent, other than paraffin, be used to lubricate, or clean, the chain. These too may damage the O-rings.

5   If the chain is of the original endless type, it is necessary to remove the complete swinging arm assembly in order to detach the chain for renewal. Refer to Chapter 4, Section 9.

6   To check if the chain is due for renewal, lay it lengthwise in a straight line and compress it endwise until all play is taken up. Anchor one end, then pull in the opposite direction to take up the play which has developed. If the chain extends by more than $\frac{1}{4}$ inch per foot, it should be renewed with the sprockets.

20 Tyres: removal and replacement

1   At some time or other the need will arise to remove and replace the tyres, either as a result of a puncture or because replacements are necessary as a result of wear. To the inexperienced, tyre changing represents a formidable task, yet if a few simple rules are observed and the technique learned, the whole operation is surprisingly simple.

2   To remove the tyre from either wheel, first detach the wheel from the machine. Deflate the tyre by removing the valve insert and when it is fully deflated, push the bead from the tyre away from the wheel rim on both sides so that the bead enters the centre well of the rim. Remove the locking cap and push the tyre valve into the tyre itself.

3   Insert a tyre lever close to the valve and lever the edge of the tyre over the outside of the wheel rim. Very little force should be necessary; if resistance is encountered it is probably due to the fact that the tyre beads have not entered the well of the wheel rim all the way round the tyre.

4   Once the tyre has been edged over the wheel rim, it is easy to work around the wheel rim so that the tyre is completely free on one side. At this stage, the inner tube can be removed.

5   Working from the other side of the wheel, ease the other edge of the tyre over the outside of the wheel rim that is furthest away. Continue to work around the rim until the tyre is free completely from the rim.

6   If a puncture has necessitated the removal of the tyre reinflate the inner tube and immerse in a bowl of water to trace the source of the leak. Mark its position and deflate the tube. Dry the tube and clean the area around the puncture with a petrol soaked rag. When the surface has dried, apply rubber solution and allow this to dry before removing the backing from the patch and applying the patch to the surface.

7   It is best to use a patch of self-vulcanising type, which will form a very permanent repair. Note that it may be necessary to remove a protective covering from the top surface of the patch, after it has sealed into position. Inner tubes made from synthetic rubber may require a special type of patch and adhesive, if a satisfactory bond is to be achieved.

8   Before refitting the tyre, check the inside to make sure that the agent which caused the puncture is not trapped. Check the outside of the tyre, particularly the tread area, to make sure nothing is trapped that may cause a further puncture.

9   If the inner tube has been patched on a number of past occasions, or if there is a tear or large hole, it is preferable to discard it and fit a new one. Sudden deflation may cause an accident, particularly if it occurs with the front wheel.

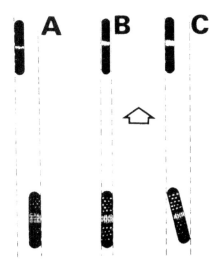

Fig. 5.7 Checking wheel alignment

A and C incorrect
B correct

10 To replace the tyre, inflate the inner tube sufficiently for it to assume a circular shape but only just. Then push it into the tyre so that it is enclosed completely. Lay the tyre on the wheel at an angle and insert the valve through the rim tape and the hole in the wheel rim. Attach the locking cap on the first few threads, sufficient to hold the valve captive in its correct location.

11 Starting at the point furthest from the valve, push the tyre bead over the edge of the wheel rim until it is located in the central well. Continue to work around the tyre in this fashion until the whole of one side of the tyre is on the rim. It may be necessary to use a tyre lever during the final stages.

12 Make sure that there is no pull on the tyre valve and again commencing with the area furthest from the valve, ease the other bead of the tyre over the edge of the rim. Finish with the area close to the valve, pushing the valve up into the tyre until the locking cap touches the rim. This will ensure the inner tube is not trapped when the last section of the bead is edged over the rim with a tyre lever.

13 Check that the inner tube is not trapped at any point. Reinflate the inner tube, and check that the tyre is seating correctly around the wheel rim. There should be a thin rib moulded around the wall of the tyre on both sides, which should be equidistant from the wheel rim at all points. If the tyre is unevenly located on the rim, try bouncing the wheel when the tyre is at the recommended pressure. It is probable that one of the beads has not pulled clear of the centre well.

14 Always run the tyres at the recommended pressures and never under or over-inflate. The correct pressures are given in the Specifications Section of this Chapter.

15 Tyre replacement is aided by dusting the side walls, particularly in the vicinity of the beads, with a liberal coating of french chalk. Washing-up liquid can also be used to good effect, but this has the disadvantage of causing the inner surfaces of the wheel rim to rust.

16 On the wire-spoked wheel models, never replace the inner tube without the rim tape in position. If this precaution is overlooked there is good chance of the ends of the spoke nipples chafing the inner tube and causing a crop of punctures.

17 Never fit a tyre that has a damaged tread or side walls. Apart from the legal aspects, there is a very great risk of a blowout, which can have serious consequences on any two-wheel vehicle.

18 Tyre valves rarely give trouble, but it is always advisable to check whether the valve itself is leaking before removing the tyre. Do not forget to fit the dust cap, which forms an effective second seal.

21 Tyre valve dust caps

1 Tyre valve dust caps are often left off when a tyre has been replaced, despite the fact that they serve an important two-fold function. Firstly they prevent dirt or other foreign matter from entering the valve and causing the valve to stick open when the tyre pump is next applied. Secondly, they form an effective second seal so that in the event of the tyre valve sticking, air will not be lost.

2 Isolated cases of sudden deflation at high speeds have been traced to the omission of the dust cap. Centrifrugal force has tended to lift the tyre valve off its seating and because the dust cap is missing, there has been no second seal. Racing innertubes contain provision for this happening because the valve inserts are fitted with stronger springs, but standard inner tubes do not, hence the need for the dust cap.

3 Note that when a dust cap is fitted for the first time, the wheel may have to be rebalanced.

22 Front wheel: balancing

1 It is customary on all high performance machines to balance the front wheel complete with tyre and tube. The out of balance forces which exist are eliminated and the handling of the machine is improved in consequence. A wheel which is badly out of balance produces through the steering a most unpleasant hammering effect at high speeds.

2 Some tyres have a balance mark on the sidewall, usually in the form of a coloured dot. This mark must be in line with the tyre valve, when the tyre is fitted to the inner tube. Even then, the wheel may require the addition of balance weights, to offset the weight of the tyre valve itself.

3 If the front wheel is raised clear of the ground and is spun, it will probably come to rest with the tyre valve or the heaviest part downward and will always come to rest in the same position. Balance weights must be added to a point diametrically opposite this heavy spot until the wheel will come to rest in ANY position after it is spun.

4 With the spoked wheels on the USA market C and HC models, balance weights which clip around the wheel spokes are available from Suzuki in 20 and 30 gramme sizes. The 20 gramme (0.04 lbs) weight is Part No 55411 – 11000, and the 30 gramme (0.07 lbs) weight has the Part No 55412 – 11000. If these are not available, wire solder, wrapped around the spokes close to the spoke nipples, forms a good substitute.

5 It is also possible to finely balance the cast-alloy wheels fitted to the rest of the GS1000 range. There are balance weights available in 20 and 30 gramme sizes, which clip to the rim of the wheel. The relevant Suzuki part Nos are 55411 – 47001 (20 gramme) and 55412 – 47001 (30 gramme).

6 There is no necessity to balance the rear wheel under normal road conditions, although it is advisable to replace the rear wheel tyre so that any balance mark is in line with the tyre valve.

23 Fault diagnosis: wheels, brakes and tyres

Symptom	Cause	Remedy
Handlebars oscillate at low speed	Buckle or flat in wheel rim, most probably front wheel (spoked wheels) Tyres not straight on rim	Check rim alignment by spinning the wheel. Correct by retensioning spokes or having wheel rebuilt on new rim. Check tyre alignment.
Machine lacks power and accelerates poorly	Brakes binding due to wrongly adjuster caliper	Check brake pads and whether piston(s) are sticking. Readjust caliper.
	Faulty caliper, on disc brake Warped disc	Replace with a new caliper. Replace disc if beyond skimming limit.
Brake squeal	Glazed pads	Lightly sand the pads, and use the brake gently for a hundred miles or so until they have a chance to bed in properly.
	Extremely dirty and dusty front brake caliper and disc assembly	Clean with water; do not use high pressure spray equipment.
Excessive lever travel on front brake	Air in system, or leak in master cylinder or caliper; worn disc pads	Bleed the brake. Renew the cylinder seals. Renew the pads.

Chapter 6 Electrical system

Contents

Specifications

Battery

Make	Yuasa or Furukawa
Type	12N14-3A
Voltage	12 volts
Capacity	14 Ah
Earth	Negative

Alternator

Make	Nippon Denso or Kokusan
Type	Permanent magnet rotor, 18-coil stator
Output	250 watts @ 5000 rpm
No load voltage	More than 16 V at 5000 rpm

Starter motor

Make	Nippon Denso or Mitsuba
Brush length	12–13 mm (0.47–0.51 in)
Service limit	6 mm (0.24 in)
Commutator undercut	0.6 mm (0.02 in)
Service limit	0.2 mm (0.008 in)

Bulbs

Headlamp	QH 60/55W (UK) All models
	QH 60/55W (USA) EC, EN, L, S models
	Sealed beam 50/35W (USA) C, and N models
Pilot lamp (UK only)	4 W
Tail/stop lamp	8/23 W (3/32 cp)
Flashing indicator lamps	23 W (32 cp)
Licence plate lamp	8 W (3 cp)
Instrument lights	3.4 W
Additional instrument lights	3.4 W (S model only)
Neutral indicator light	3.4 W
Oil pressure warning light	3.4 W
High beam indicator light	3.4 W
Indicator warning light	3.4 W
Digital gear position lights (5)	3.4 W (L model only)

All bulbs rated at 12 volt.

1 General description

The Suzuki GS1000 models are fitted with a 12 volt electrical system, powered by an alternator mounted on the extreme left-hand end of the crankshaft. The alternator, which produces alternating current, is the three-stage type, having a permanent magnet rotor and an eighteen-coil stator.

During day-light running, other than in the USA, no lights are in permanent operation. This means only two of the three output stages are utilised. The third stage remains out of circuit until the lighting switch is operated, when additional current is required to meet the demands of the lighting circuit.

The AC current produced by the alternator is converted into DC current by a full-wave silicon rectifier and is controlled to meet the voltage demands of the system by a solid state regulator (SCR).

2 Testing the electrical system: general

1   Checking the electrical output and the performance of the various components within the charging system requires the use of test equipment of the multi-meter type. When carrying out checks, care must be taken to follow the procedures laid down and so prevent inadvertent incorrect connections or short circuits. Irreparable damage to individual components may result if reversal of current or shorting occurs. It is advised that unless some previous experience has been gained in auto-electrical testing, the machine be returned to a Suzuki Service Agent or auto-electrician, who will be qualified to carry out the work and have the necessary test equipment.

2   If the performance of the charging system is suspect, the system as a whole should be checked first, followed by testing of the individual components to isolate the fault. The three main components are the alternator, the rectifier, and the regulator. Before commencing the tests, ensure that the battery is fully charged, as described in Section 7.

3 Charging system: checking the output

1st test
1   The first test is performed in a no-load state, with the regulator disconnected, and will verify whether the alternator and rectifier are functioning. Raise the dualseat so that the access may be made to the wiring harness.

2   Disconnect the yellow wire running from the regulator. Disconnect the green with white tracer wire running from the alternator, and the white with red tracer wire from the rectifier. Connect these latter two wires together to by-pass the lighting switch. Turn the lighting switch off. Connect a 0–20 v DC voltmeter across the battery terminals. Start the engine and increase the engine speed to 5000 rpm. At this speed the indicated voltage should be 14–15 V. If the voltage reading is above 15 V, the regulator is faulty. If the voltage indicated is below 14 V, the alternator or rectifier is faulty. To eliminate which of the two is the faulty component, refer to Section 4 or 5 of this Chapter.

2nd test
3   A second test should be performed to confirm the regulator is, indeed, not functioning correctly. Reconnect the wires so that they are restored to the correct original positions. Turn the lighting switch off, so that only two of the three output stages are in circuit. On models supplied to the USA, where the lighting switch is locked in the 'On' position, remove the switch knob and manually turn the switch off.

4   Start the engine and again check the voltage at 5000 rpm. If the indicated voltage is within the 14–15 V range, the regulator is functioning correctly. A reading above or below this range indicates that the regulator is faulty. Before consigning the regulator to the scrap bin, check that all the wiring connec-

tions are clean and tight. Also check that the battery is fully charged. Note that the battery must be fully up to its 12 V capacity before attempting tests on any of these three main components. False readings will occur if the battery is not functioning correctly in any way. Repeat the second test with these checks carried out. If the readings still indicate an incorrectly functioning regulator, it must be replaced; there are no serviceable parts in the regulator unit. A replacement regulator must be of the same make as the alternator.

4 Rectifier: location and testing

1   The rectifier is fitted behind the left-hand side cover. If the rectifier is suspected of being faulty after carrying out test No 1 in the preceding Section, it may be tested in situ using a multi-meter set to the resistance function.

2   Disconnect the battery to isolate the electrical system and then disconnect all five wires which lead to the rectifier unit. Connect the negative ohmmeter lead to the regulator earth terminal (black/white) and then test the continuity between the earth and the following terminals: yellow, white/red, white/blue, and red. Continuity should be indicated on each. Reverse the polarity of the ohmmeter and repeat the test. No continuity should be indicated. Carry out the same series of tests but with the output terminal (red) used as the common test terminal, in place of the earth terminal. The correct results should be the reverse of those given for the first part of the test. If one or more incorrect readings is found, the rectifier must be renewed.

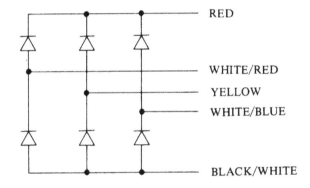

Fig. 6.1 Rectifier circuit

5 Alternator: testing

1   If after carrying out test No 1 in Section 3 of this Chapter it was found that the alternator or rectifier was not functioning correctly, as indicated by the voltage reading, the alternator may be tested after removal of the stator from the machine, using a multi-meter set to the resistance function.

2   Disconnect the leads running from the alternator and check the continuity between the three wires, making the test in pairs. The correct resistance is 0.65 ohms ± 0.05%. If there is found to be no continuity between any two of the wires, or if the resistance is too low, a short-circuit or open-circuit is evident, and the stator must be renewed. Two makes of alternator are utilised, Denso and Kokusan. When renewing the stator, ensure that the replacement component is of the same manufacture as the rotor.

3   Check the resistance between each wire and the stator core. No continuity should be found.

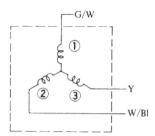

Fig. 6.2 Alternator circuit

---

6   Battery: examination and maintenance

1   The GS1000 range is fitted with a 12 volt battery with a 14 ampere hour capacity.

2   The transparent plastic case of the battery permits the upper and lower levels of the electrolyte to be observed without disturbing the battery by removing the left-hand side cover. Maintenance is normally limited to keeping the electrolyte level between the prescribed upper and lower limits and making sure that the vent is not blocked. The lead plates and their separators are also visible through the transparent case, a further guide to the general condition of the battery.

3   Unless acid is spilt, as may occur if the machine falls over, the electrolyte should always be topped up with distilled water to restore the correct level. If acid is spilt onto any part of the machine, it should be neutralised with an alkali such as washing soda or baking powder and washed away with plenty of water, otherwise serious corrosion will occur. Top up- with sulphuric acid of the correct specific gravity (1.260 to 1.280) only when spillage has occurred. Check that the vent pipe is well clear of the frame or any of the other cycle parts.

4   It is seldom practicable to repair a cracked battery case because the acid present in the joint will prevent the formation of an effective seal. It is always best to renew a cracked battery, especially in view of the corrosion which will be caused if the acid continues to leak.

5   If the machine is not used for a period, it is advisable to remove the battery and give it a refresher charge every six weeks or so from a battery charger. If the battery is permitted to discharge completely, the plates will sulphate and render the battery useless.

6   Occasionally, check the condition of the battery terminals to ensure that corrosion is not taking place and that the electrical connections are tight. If corrosion has occurred, it should be cleaned away by scraping with a knife and then using emery cloth to remove the final traces. Remake the electrical connections whilst the joint is still clean, then smear the assembly with petroleum jelly (NOT grease) to prevent recurrence of the corrosion. Badly corroded connections can have a high electrical resistance and may give the impression of a complete battery failure.

---

7   Battery: charging procedure

1   The normal charging rate for batteries of up to 14 amp. hour capacity is $1\frac{1}{2}$ amps. It is permissible to charge at a more rapid rate in an emergency but this shortens the life of the battery, and should be avoided. Always remove the vent caps when recharging a battery, otherwise the gas created within the battery when charging takes place will explode and burst the case with disastrous consequences.

2   When the battery is reconnected to the machine, it is important that the two leads are replaced on the correct terminals. If the leads are inadvertently reversed, the electrical system will be damaged permanently. The rectifier will be destroyed by a reversal of the current flow.

6.2 Electrolyte level must be kept between upper and lower level lines

---

8   Fuses: location and replacement

1   A bank of fuses is contained within a small plastic box located near the regulator and sharing the same mounting bracket. To gain access to the fuse box, removal of the left-hand side frame cover is necessary. The front section of the fuse box pulls away from the base section leaving the four fuses visible. The inside of the fuse box front section contains two spare fuses, one of each rating used in the system. The fuses used are three rated at 10A, and one at 15A. The main fuse, 15A, protects all the electrical systems. This is situated at the top of the fuse box. Below is the 10A fuse protecting the headlamp, tail lamp, rear number plate lamp, instrument lamps, and high beam indicator lamp. The second 10 A fuse protects the brake lamp, indicators and the warning lamp repeater, the horn and the indicator warning buzzer (where fitted). The third 10A fuse in the lower position protects the electric start system and the ignition system.

2   Before replacing a fuse that has blown, check for the cause of the short. Check that no obvious short circuit has occured. This will involve checking the electrical circuit to correct the fault. If this rule is not observed, the fuse will almost certainly blow again.

3   When a fuse blows while the machine is in operation, and no spare is available, a 'get you home' remedy is to remove the blown fuse and wrap it in silver paper before replacing it in the fuseholder. The silver paper will restore the electrical continuity by bridging the broken fuse wire. This expedient should NEVER be used if there is evidence of a short circuit or other major electrical fault, otherwise more serious damage will be caused. Replace the 'doctored' fuse at the earliest possible opportunity, to restore full circuit protection.

---

9   Starter motor: removal, examination and replacement

1   An electric starter motor, operated from a small push-button on the right-hand side of the handlebars, is the method provided for starting the engine. The starter motor is mounted within a compartment at the rear of the cylinder block, closed by an oblong, chromium plated cover. Current is supplied from the battery via a heavy duty solenoid switch and a cable capable of carrying the very high current demanded by the starter motor on the initial start-up.

2   The starter motor drives a free running clutch immediately behind the generator rotor. The clutch ensures the starter motor

drive is disconnected from the primary transmission immediately the engine starts. It operates on the roller and ramp principle; as the starter driven pinion is rotated the rollers are forced into wedge shaped slots in the clutch hub causing the pinion boss to lock with the clutch hub. Once the engine has started the rollers are freed and drive is lost.

3   To remove the starter motor from the engine unit, first disconnect the positive lead from the battery, to isolate the electrical system. Remove the cover plate which encloses the starter motor and detach the heavy duty cable from the terminal on the starter motor body. Temporarily detach the oil pressure switch lead from the switch. The starter motor is secured to the crankcase by two bolts which pass through the left-hand end of the motor casting. When these bolts are withdrawn, the motor can be prised out of position and lifted out of its compartment.

4   The parts of the starter motor most likely to require attention are the brushes. The end cover is retained by the two long screws which pass through the lugs cast on both end pieces. If the screws are withdrawn, the end cover can be lifted away and the brush gear exposed.

5   Lift up the spring clips which bear on the end of each brush and remove the brushes from their holders. The standard length and wear limits of the brushes are the same for each make of starter employed. They are as follows:

standard length	service limit
12–13 mm	6 mm
(0.47–0.51 in)	(0.24 in)

6   When the brushes are badly worn, the starter motor will be unable to produce sufficient power to turn the engine over to facilitate starting.

7   Before the brushes are replaced, make sure the commutator on which they bear is clean. Clean with a strip of fine glass paper cloth pressed against the commutator whilst the latter is revolved by hand. Emery paper should NOT be used because abrasive fragments may embed themselves in the soft metal of the commutator and cause excessive wear of the brushes. Finish off the commutator with metal polish to give a smooth surface and finally wipe the segments over with a methylated spirits soaked rag to ensure a grease free surface. Check that the mica insulators, which lie between the segments of the commutator, are undercut. The standard groove depth is 0.6 mm (0.02 in) but if the average groove depth is less than 0.2 mm (0.008 in) the armature should be renewed or returned to a Suzuki Service Agent for re-cutting.

8   Replace the brushes in their holders and check that they slide quite freely. Make sure the brushes are replaced in their original positions because they will have worn to the profile of the commutator. Replace and tighten the end cover, then replace the starter motor and cable in the housing, tighten down and re-make the electrical connection to the solenoid switch. Check that the starter motor functions correctly before replacing the compartment cover and sealing gasket.

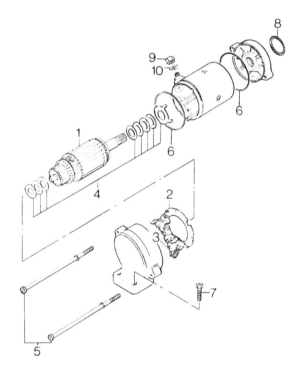

Fig. 6.3 Starter motor assembly

1	Armature	6	O-ring – 2 off
2	Brush holder plate	7	Bolt – 2 off
3	Brush set	8	O-ring
4	Shim set	9	Nut
5	Bolt – 2 off	10	Spring washer

10 Starter motor free running clutch: construction and renovation

1   Although a mechanical and not an electrical component, it is appropriate to include the free running clutch in this Chapter because it is an essential part of the electric starter system.

2   As mentioned in Chapter 1, the free running clutch is built into the alternator rotor assembly and will be found in the back of the rotor when the latter is removed from the left-hand end of the crankshaft. The only parts likely to require attention are the rollers and their springs, or the bush in the centre of the driven sprocket. Access to the rollers is gained by removing the three Allen head bolts which retain the clutch body to the rear of the alternator rotor. Signs of wear or damage will be obvious and will necessitate renewal of the worn or damaged parts.

3   The bush in the centre of the driven sprocket behind the clutch will need renewal only after very extensive service.

4   To check whether the clutch is operating correctly, turn the driven sprocket anticlockwise. This should force the spring loaded rollers against the clutch hub and cause it to tighten on the hub as the drive is taken up.

5   If the starter clutch has been dismantled, make sure the three Allen head bolts are tightened fully and a small quantity of locking fluid applied to the bolt threads, to prevent their working loose.

9.1 Starter motor situated under a removable chromed cover

10.2a Remove the Allen head bolts to separate the clutch body from rear of rotor and ...

10.2b ... lift clutch unit away having first ...

10.2c ... removed the primary drive sprocket from behind the rotor

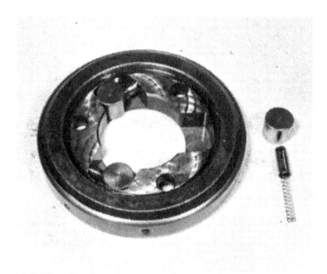

10.2d Examine the rollers and their springs for wear

10.2e Note the plate between the clutch and rotor

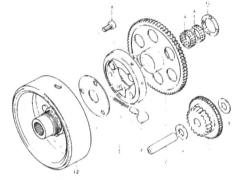

Fig. 6.4 Starter clutch assembly

1	Idler gear	8	Allen screw – 3 off
2	Spindle	9	Needle roller bearing –
3	Shim – 2 off		2 off
4	Clutch assembly	10	Washer
5	Roller – 3 off	11	Backing plate
6	Spring – 3 off	12	Alternator rotor
7	Plunger – 3 off		

11 Starter solenoid switch: function and location

1   The starter motor switch is designed to work on the electro-magnetic principle. When the starter motor button is depressed, current from the battery passes through windings in the switch solenoid and generates an electro-magnetic force which causes a set of contact points to close. Immediately the points close, the starter motor is energised and a very heavy current is drawn from the battery.

2   This arrangement is used for at least two reasons. Firstly, the starter motor current is drawn only when the button is depressed and is cut off again when pressure on the button is released. This ensures minimum drainage on the battery. Secondly, if the battery is in a low state of charge, there will not be sufficient current to cause the solenoid contacts to close. In consequence, it is not possible to place an excessive drain on the battery which, in some circumstances, can cause the plates to overheat and shed their coatings. If the starter will not operate, first suspect a discharged battery. This can be checked by trying the horn or switching on the lights. If this check shows the battery to be in good shape, suspect the starter switch which should come into action with a pronounced click. It is located under the dualseat, close to the battery, and can be identified by the heavy duty starter cable connected to it. It is not possible to effect a satisfactory repair if the switch malfunctions; it must be renewed.

12 Headlamp: replacing bulbs and adjusting beam height

1   In order to gain access to the headlamp bulbs it is necessary to first remove the rim, complete with the reflector and headlamp glass. The rim is retained by three crosshead screws equally spaced around the headlamp shell. Remove the screws completely and draw the rim from the headlamp shell.

2   UK models have a main headlamp bulb which is a push fit into the central bulb holder of the reflector. The bulb fitted to all the UK models is of the quartz halogen type, and has a 60/55W rating. The locating lugs on the bulb ensure that it is always correctly positioned, and, therefore, correctly focussed, by engaging with slots on the outside of the central bulb holder of the reflector. The bulb is retained in these slots by a retaining spring and rubber boot. When removing the bulb, on no account should the quartz envelope, be handled. Use a piece of clean, dry cloth to insulate the bulb from the hand when removal, or insertion, is necessary. If the glass is touched by the skin, a 'hot spot' will be formed. Due to the high temperatures developed during the working of these bulbs, this will result in a weak point, and the bulb will fail prematurely. The pilot lamp bulb is a bayonet fitting and fits within a bulb holder which has the same type of attachment as the main bulb to the reflector. This bulb has a rating of 6W.

3   US models have either a sealed beam headlamp unit, rated at 50/40W, or the quartz halogen 60/55W unit, as fitted to the UK versions, depending on the model. The EC, EN, L, and S versions all have the quartz halogen unit. This unit is identical to the ones fitted to the UK models, with the exception that no provision for a pilot lamp is made. The models with the standard sealed beam headlamp unit, are also not equipped with parking lamps. For information on the quartz halogen unit, refer to the previous paragraph of this section. The headlamp unit of the GS1000L model is slightly different from those fitted to the rest of the range. The chromed unit appears initially to be a one-piece component. Removing two crosshead screws, one each side of the shell, however, allows the rim and headlamp internal unit to pull free from the shell. Rolling back the rubber boot reveals a retaining spring fitted to the rear of the bulb. Free the spring from the locating slot on the outside edge of the central bulb fitting hole, in the back of the reflector. Releasing the

spring from tension enables the bulb to be removed. Refer to the previous section for information on the quartz halogen bulb fitted to this headlamp.

With the sealed beam unit, if one filament blows, the complete unit must be renewed. To release the lamp unit, remove the horizontal adjusting screw and the three retaining screws and washers, from the collar that clamps the light shell to the headlamp rim. Make a note of the setting of the adjusting screw, otherwise it will be necessary to re-adjust the beam height after installing the new light unit. Remove two screws to separate the sealed beam headlamp unit and the retaining ring, from the rim, and remove a further three screws to release the sealed headlamp unit from the retaining ring. To reassemble the headlamp, reverse the dismantling procedure, noting that the marking TOP, faces upwards.

4   Beam angle is adjusted by turning the adjusting screw fitted in the nine o'clock position, when the headlamp is viewed from the front. Turning the adjusting screw anti-clockwise, moves the beam towards the right, and turning the adjuster clockwise, results in the beam being redirected towards the left. The vertical beam height is adjusted manually. Loosen the left-hand and right-hand headlamp shell mounting bolts, and raise or lower the complete unit until the correct setting is achieved. Retighten the mounting bolts when height is correct. To adjust the horizontal beam angle on the GS1000 L model, a small crosshead screw is provided. This is fitted to the right-hand side of the headlamp rim at approximately the four o'clock position when the machine is viewed from a front right three-quarters angle. Turn the adjusting screw clockwise or anticlockwise in the same manner as the on the other models. Vertical adjustment is as for the other models.

5   UK lighting regulations stipulate that the lighting system must be arranged so that the light will not dazzle a person standing at a distance greater than 25 feet from the lamp, whose eye level is not less than 3 feet 6 inches above that plane. It is easy to approximate this setting by placing the machine 25 feet away from a wall, on a level road, and setting the beam height so that it is concentrated at the same height as the distance of the centre of the headlamp from the ground. The rider must be seated normally during this operation and also the pillion passenger, if one is carried regularly.

13 Stop and tail lamp: replacing the bulbs

1   The tail lamp unit of all the GS1000 range contains two bulbs. The larger of the two bulbs is of the twin filament type with a rating of 8/23W (3/32 cp). The higher rated filament operates to give visible warning when the brake is applied, and the lower rated filament acts as a permanent warning lamp with the lights switched on. The second, single filament 8W (3 cp), bulb is used to illuminate the rear number plate. Stop lamp switches operate in conjunction with both the front and rear brakes in order that the system meets the statutory requirements of all the states and countries to which the machines are exported.

2   To gain access to the bulbs, unscrew the four crosshead screws which retain the plastic lens cover in position. The bulbs both have a bayonet fitting, and the tail/stop lamp bulb has offset pins. This is in order that the stop lamp filament cannot be inadvertently connected with the tail lamp and vice versa.

14 Flashing indicator lamps: replacing bulbs

1   Flashing indicator lamps are fitted to the front and rear of the machine. They are mounted on short stalks through which the wires pass. Access to each bulb is gained by removing the two screws holding the plastic lens cover. The bulbs are of 23W (32 cp) rating, and are retained by a bayonet fitting.

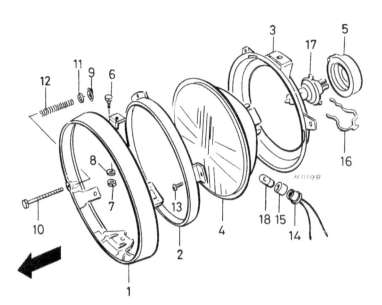

1 Headlamp rim
2 Outer retaining rim
3 Mounting ring
4 Reflector unit
5 Rubber boot
6 Screw – 2 off
7 Nut – 2 off
8 Spacer – 2 off
9 Nut
10 Adjusting screw
11 Washer
12 Spring
13 Screw – 2 off
14 Bulb holder
15 Grommet
16 Spring clip
17 Main bulb
18 Pilot bulb

Fig. 6.5 Headlamp assembly Q.H.

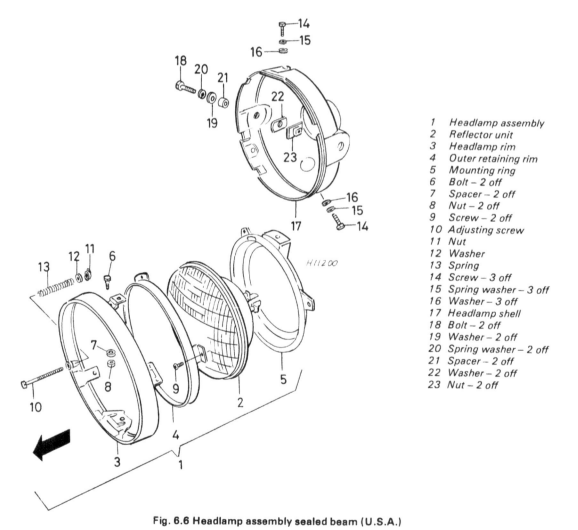

1 Headlamp assembly
2 Reflector unit
3 Headlamp rim
4 Outer retaining rim
5 Mounting ring
6 Bolt – 2 off
7 Spacer – 2 off
8 Nut – 2 off
9 Screw – 2 off
10 Adjusting screw
11 Nut
12 Washer
13 Spring
14 Screw – 3 off
15 Spring washer – 3 off
16 Washer – 3 off
17 Headlamp shell
18 Bolt – 2 off
19 Washer – 2 off
20 Spring washer – 2 off
21 Spacer – 2 off
22 Washer – 2 off
23 Nut – 2 off

Fig. 6.6 Headlamp assembly sealed beam (U.S.A.)

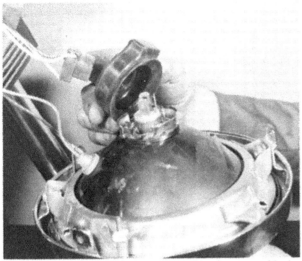

12.2a Main bulb holder is secured by a rubber boot, note block connector separated

12.2b Take care when handling quartz halogen bulb

12.2c Pilot bulb holder has bayonet fitting as has the bulb

12.2d Three screws held reflector/rim unit to headlamp shell

12.4 Use adjusting screw to maintain correct beam adjustment

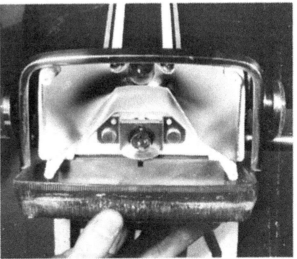

13.2 Stop/tail lamp lens is held by four screws

14.1 Indicator bulbs have a bayonet fixing

15 Flasher unit: location and replacement

1   The flasher relay unit is located behind the left-hand side cover and is supported on anti-vibration mountings made of rubber, attached to the base plate.
2   If the flasher unit is functioning correctly, a series of audible clicks will be heard when the indicator lamps are in operation. If the unit malfunctions and all the bulbs are in working order, the usual symptom is one initial flash before the unit goes dead; it will be necessary to replace the unit complete if the fault cannot be attributed to any other cause.
3   In addition to the flasher unit, either an audible warning buzzer, or self-cancelling unit, is fitted, depending on the particular model.
4   The audible warning device is fitted below the petrol tank, suspended from a frame cross-piece, between the main frame top tubes. If the device malfunctions it must be replaced as a complete unit.
5   The electronic self-cancelling unit is incorporated in the indicator system. The unit automatically turns the indicators off, a certain time after the indicator switch has been operated. The time lapse is dependent on the speed of the machine. If the machine is travelling faster than 15 km/h (9 mph), the unit cancels the indicators after approximately 9 seconds. When the machine is stationary, or running at a speed of less than 15 km/h (9 mph), the unit does not come into operation; the indicators continue to function until either cancelled manually, or the machine speed increases. The system may be overridden manually, for example, when overtaking a single vehicle when the full 9 seconds of automatically controlled operation would not be required, by simply pressing the switch downwards.
6   Take great care when handling either unit because they are easily damaged if dropped or subjected to rough treatment.

16 Speedometer and tachometer: replacement of bulbs

1   Bulbs fitted to each instrument illuminate the dials when the main lights are in operation. The four bulbs fitted to the two instruments all have the same type of push fit bulb holder and are all of the small bayonet type fitting.
2   Access to the bulbs and holders is gained by removing the nuts from the studs on the bottom of each instrument which retain the chromed cover to each. This cannot be done, however, until the main shared base plate is separated from the lower instrument covers. Disconnect the drive cables and lift each instrument up and pull out the bulb holders.

17 Supplementary instrumentation: replacement of bulbs GS1000 S model

1   The GS 1000 S model is fitted with supplementary instrumentation in the form of three extra gauges, inserted into the upper, forward section, of the small handlebar mounted fairing. The gauges inform the operator of the temperature of the engine oil and the amount of petrol in the tank; the third dial is a clock. Each of these instruments is illuminated, in conjunction with the speedometer and tachometer, by a single bayonet fitting bulb below each dial.
2   The easiest way to facilitate access to the bulbs and holders is by removing the fairing. With this detached, the instrument console lower cover can be removed and lowered. Disconnect the various gauge leads and pull out the bulb holders whilst holding up the console.

18 Indicator panel lamps: replacement of bulbs

1   An indicator lamp panel, advising the operator of the functioning of certain components, is fitted between the speedometer and tachometer heads. The panel contains four warning bulbs on all the models except the GS1000 S which contains five. To gain access to the bulbs remove the four screws which pass through the upper cover and lift the cover away. Each bulb is fitted to a separate bulb holder.

19 Gear selector indicator: location and operation GS1000 L model

1   A gear selection indicator is fitted to the GS1000 L models, in addition to the neutral warning light. In this system, a switch on the change drum illuminates a digital display unit fitted in a console mounted cross-wise between the speedometer and tachometer heads. The display unit indicates which of the gears in turn has been selected, by illuminating separate bulbs behind a numbered panel. If the unit malfunctions, removal of the warning lamp console cover will facilitate inspection of the bulbs and their holders. Each bulb can be replaced individually.

20 Horn: location and examination

1   The horn on all the GS1000 models consists of two separate units each flexibly mounted on a steel strip bolted to the frame below the petrol tank. The steel mounting strip is also suspended by flexible rubber mounts. With both the two horn units and their individual mounting strips being rubber mounted, a certain amount of flex will be evident even when the retaining bolts are fully tightened. Care should be taken when refitting the horn units to ensure that they are positioned correctly, and do not come into contact with the underside of the petrol tank. When this occurs, with the machine in operation, a drumming sound will be produced, and a slight vibration may be felt through the petrol tank sides.
2   If one, or both, of the horn units manfunction, they must be replaced, it is a statutory requirement that the machine must be fitted with a horn in working order.

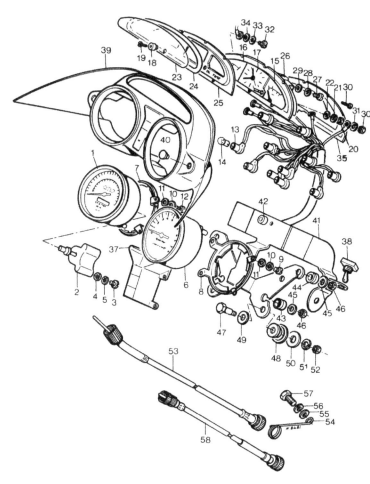

Fig. 6.7 Speedometer and tachometer heads

13 Bulb holder unit
14 Bulb – 11 off
15 Oil pressure gauge
16 Fuel gauge
17 Clock
18 Clock adjusting knob
19 Screw
20 Nut – 8 off
21 Spring washer – 8 off
22 Washer – 8 off
23 Instrument console glass
24 Sealing ring
25 Upper instrument console cover
26 Instrument console
27 Screw – 8 off
28 Spring washer – 8 off
29 Washer – 8 off
30 Nut – 7 off
31 Spring washer – 7 off
32 Screw
33 Spring washer
34 Washer
35 Lower instrument console cover
36 Screw – 4 off
37 Warning instrument console
38 Clock switch
39 Instrument housing
40 Grommet
41 Instrument assembly mounting bracket
42 Grommet
43 Rear mounting grommet – 2 off
44 Front mounting grommet – 2 off
45 Washer – 4 off
46 Nut – 4 off
47 Bolt – 2 off
48 Grommet – 2 off
49 Washer
50 Washer
51 Spring washer – 2 off
52 Nut – 2 off
53 Speedometer drive cable
54 Cable guide
55 Washer
56 Spring washer
57 Bolt
58 Tachometer drive cable

1 Speedometer assembly
2 Trip reset button and holder
3 Screw – 2 off
4 Washer – 2 off
5 Spring washer – 2 off
6 Tachometer assembly
7 Speedometer retainer
8 Tachometer retainer
9 Screw – 4 off
10 Spring washer – 8 off
11 Washer – 8 off
12 Screw – 4 off

21 Ignition switch: removal and replacement

1 The combined ignition, lighting master switch and steering lock is mounted in a separate unit below the warning lamp console forward of the handlebar cross-piece.
2 If the switch proves defective, it may be removed after first removing the plastic switch cover. This can be pulled upwards and displaced. Remove the two retaining screws and disconnect the lead connecting the switch to the wiring loom, at the connector.
3 Reassembly of the switch can be made in the reverse procedure as described for dismantling. Repair is rarely practicable. It is preferable to purchase a new switch unit, which will probably necessitate the use of a different key.

22 Stop lamp switch: adjustment

1 All models have a stop lamp switch fitted to operate in conjunction with the front and rear brakes. The switch for the rear brake stop lamp, working from the rear brake pedal, is located immediately to the rear of the crankcase, on the right-hand side of the machine. It has a threaded body giving a range of adjustment.
2 If the stop lamp is late in operating, slacken the locknuts and turn the body of the switch anti-clockwise so that the switch rises upwards from the bracket to which it is attached. When the adjustment seems near correct, tighten the locknuts and test.
3 If the lamp operates too early, the locknuts should be slackened and the switch body turned clockwise so that it is lowered in relation to the mounting bracket. The operation of the lamp is partly personal preference, but as a guide, it should operate after the brake pedal has been depressed by about $1\frac{1}{2}$ cm ($\frac{5}{8}$ inch).
4 A stop lamp switch is also incorporated in the front brake system. The mechanical switch is a push fit in the handlebar lever stock. If the switch 'timing' needs to be adjusted, this can be accomplished by loosening the two screws that retain the switch to the body, and repositioning as necessary. If the switch malfunctions, repair is impracticable. The switch should be renewed.

23 Handlebar switches: general

1 In general, the switches give little trouble, but if necessary, they can be dismantled by separating the halves which form a split clamp around the handlebars. Note that the machine cannot be started with the ignition cut-out, on the right-hand switch unit, in the 'OFF' position; it must be moved to the 'RUN' position.

2 The GS1000 EN, S, and L models, have, in addition to the ignition cut-out switch, been fitted with an electric starter interlock system. This device interconnects the starter and the clutch handlebar lever; unless the clutch lever is operated, to disengage the clutch, the starter motor will not come into operation. The device operates by using a switch that interrupts the current passing from the starter motor button, until the clutch is disengaged.

3 Always disconnect the battery before removing any of the switches, to prevent the possibility of a short circuit. Most troubles are caused by dirty contacts, but in the event of the breakage of some internal part, it will be necessary to renew the complete switch.

4 Because the internal components of each switch are very small, and therefore, difficult to dismantle and reassemble, it is suggested a special electrical contact cleaner be used to clean corroded contacts. This can be sprayed into each switch, without the need for dismantling.

24 Neutral indicator switch: location and removal

1 A switch is incorporated in the gearbox which indicates, via green lamp on the warning lamp console, when neutral gear position has been selected. The switch is fitted into the left-hand side gearbox wall, roughly below the final drive sprocket. In the event of failure, the switch may be unscrewed without draining the transmission oil. It is retained by two screws. Disconnect the switch lead by unscrewing the single screw. Note that the switch can easily shear if excessive force is used when tightening the unit during refitting.

25 Fuel gauge circuit: testing

1 If the fuel gauge appears to be faulty, the source of the fault is easily traced. Locate the fuel gauge sender leads. Separate the connectors, then join the harness side of the two leads with a short length of insulated wire. This eliminates the sender unit from the circuit, and if the ignition is now switched on, the fuel gauge should indicate full, indicating that the fault must lie in the sender unit. If the gauge does not respond, the fault lies in the instrument itself or its supply from the ignition switch.

2 If the fuel gauge sender is suspect, remove the fuel tank as described in Chapter 2. With the tank inverted on soft rag to protect the paintwork, remove the sender holding bolts and manoeuvre it out of the tank, taking care not to damage or bend the float. Check the sender resistances at the full and empty positions. If the results are not as shown below, the sender unit should be renewed.

Full (float fully raised) .... 0.5-5.5 ohm
Empty (float fully lowered) .... 102-118 ohm

3 The gauge is housed in the tachometer and if renewal is required it will be necessary to renew the complete tachometer unit; the gauge is not available as a separate item. If the sender unit check described above indicates that the fault lies in the gauge, check the gauge wiring as follows:

4 Locate the block connector(s) from the instrument console and identify the fuel circuit wires, yellow/black, orange and black/white (see wiring diagrams). Check for continuity between the yellow/black wire terminal of the connectors and the yellow/black wire terminal of the gauge. Also check for continuity between the black/white wire terminal of the gauge and a good earth point on the machine. In both cases low resistance (continuity) should be indicated. Check the power supply to the gauge by testing for full battery voltage at the orange wire terminal of the connector, with the ignition switched on. Do not take the reading at the gauge itself because its power supply is reduced by a voltage regulator situated in the instrument head. If these tests indicate that the supply to the gauge is correct, and the fault still persists, then the gauge requires renewal.

26 Fault diagnosis: Electrical system

Symptom	Cause	Remedy
Complete electrical failure	Blown fuse	Check wiring and electrical components for short circuit before fitting new fuse.
	Isolated battery	Check battery connections, also whether connections show signs of corrosion.
Dim lights, horn and starter inoperative	Discharged battery	Remove battery and charge with battery charger. Check generator output and voltage regulator settings.
Constantly blowing bulbs	Vibration or poor earth connection	Check security of bulb holders. Check earth return connections.
Starter motor sluggish	Worn brushes	Remove starter motor. Renew brushes.
Parking lights dim rapidly	Battery will not hold charge	Renew battery at earliest opportunity.
Flashing indicators do not operate	Blown bulb	Renew bulb.
	Damaged flasher unit	Renew flasher unit.

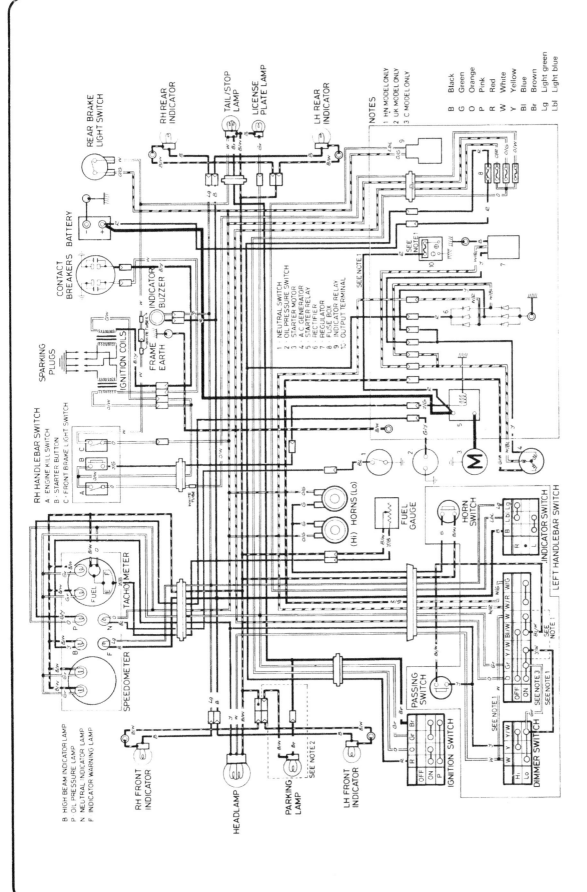

Wiring diagram – GS1000 C, HC and HN models

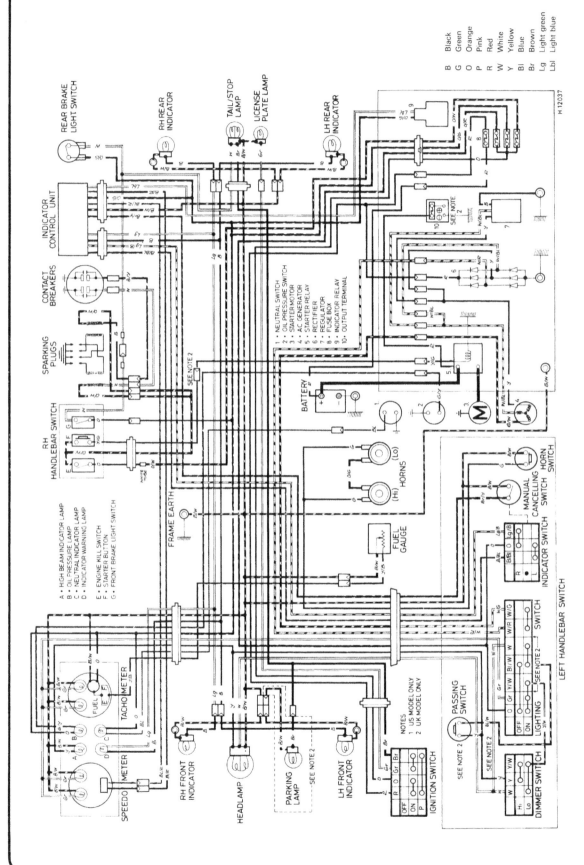

Wiring diagram – GS1000 EC and EN models

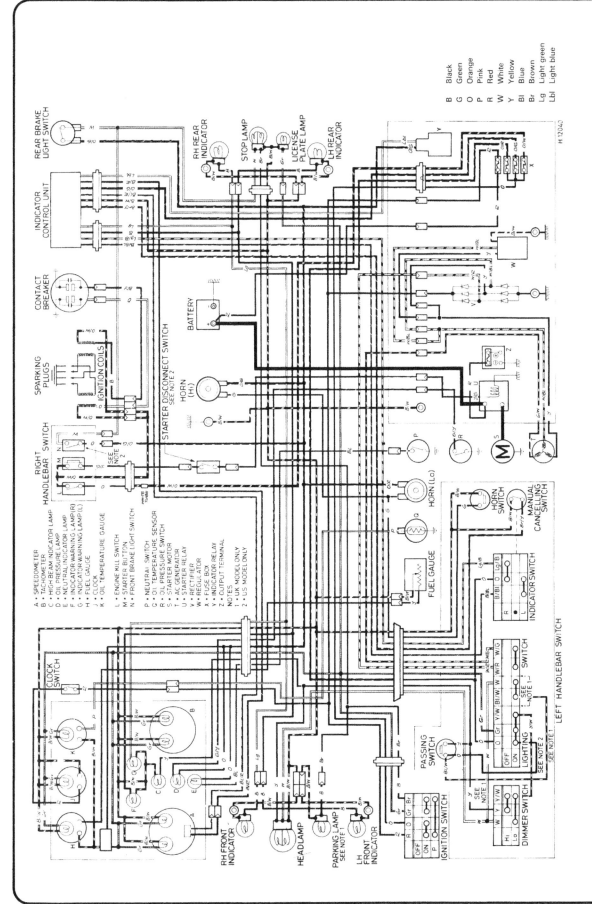

Wiring diagram – GS1000 S model

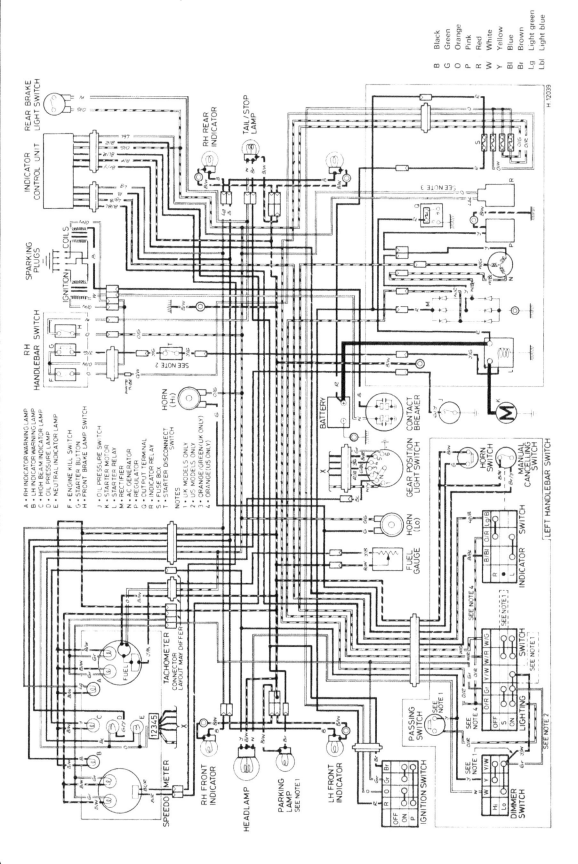

Wiring diagram GS1000 L model

English/American terminology

Because this book has been written in England, British English component names, phrases and spellings have been used throughout. American English usage is quite often different and whereas normally no confusion should occur, a list of equivalent terminology is given below.

English	American	English	American
Air filter	Air cleaner	Number plate	License plate
Alignment (headlamp)	Aim	Output or layshaft	Countershaft
Allen screw/key	Socket screw/wrench	Panniers	Side cases
Anticlockwise	Counterclockwise	Paraffin	Kerosene
Bottom/top gear	Low/high gear	Petrol	Gasoline
Bottom/top yoke	Bottom/top triple clamp	Petrol/fuel tank	Gas tank
Bush	Bushing	Pinking	Pinging
Carburettor	Carburetor	Rear suspension unit	Rear shock absorber
Catch	Latch	Rocker cover	Valve cover
Circlip	Snap ring	Selector	Shifter
Clutch drum	Clutch housing	Self-locking pliers	Vise-grips
Dip switch	Dimmer switch	Side or parking lamp	Parking or auxiliary light
Disulphide	Disulfide	Side or prop stand	Kick stand
Dynamo	DC generator	Silencer	Muffler
Earth	Ground	Spanner	Wrench
End float	End play	Split pin	Cotter pin
Engineer's blue	Machinist's dye	Stanchion	Tube
Exhaust pipe	Header	Sulphuric	Sulfuric
Fault diagnosis	Trouble shooting	Sump	Oil pan
Float chamber	Float bowl	Swinging arm	Swingarm
Footrest	Footpeg	Tab washer	Lock washer
Fuel/petrol tap	Petcock	Top box	Trunk
Gaiter	Boot	Torch	Flashlight
Gearbox	Transmission	Two/four stroke	Two/four cycle
Gearchange	Shift	Tyre	Tire
Gudgeon pin	Wrist/piston pin	Valve collar	Valve retainer
Indicator	Turn signal	Valve collets	Valve cotters
Inlet	Intake	Vice	Vise
Input shaft or mainshaft	Mainshaft	Wheel spindle	Axle
Kickstart	Kickstarter	White spirit	Stoddard solvent
Lower leg	Slider	Windscreen	Windshield
Mudguard	Fender		

Index